21st CENTURY CLUES

Front Cover Image © Keith Lehrer
<http://www.u.arizona.edu/~lehrer/ga.htm>
Reprinted By Permission

BOOK SUMMARY

21st Century Clues: Essays in Ethics, Ontology, and Time Travel

Charles Tandy, Ph.D.

These 14 previously published essays (2001-2008) take us on a journey toward a transhuman future rarely explored by professional philosophers. The journey's clues come less from the techno-optimist predictive route of many futurists – more from the disciplined context of professional philosophizing. Included is a discussion of eight types of time machines. From basic biostasis to time viewing to actual time travel, Dr. Tandy expands the categories we use to frame the future. From universe to multiverse to many-multiverses, Tandy takes us to new worlds. This allows a richness to our understanding of not only what can be, but what ought to be. Now it is time to grow up.

21st CENTURY CLUES

Essays in Ethics, Ontology, and Time Travel

by

CHARLES TANDY, PH.D.

Ria University Press

www.ria.edu/rup

2010

Printed in the United States of America

Ria University Press   Palo Alto, California

21st CENTURY CLUES

Essays in Ethics, Ontology, and Time Travel

by

Charles Tandy, Ph.D.

FIRST PUBLISHED IN HARDBACK AND PAPERBACK 2010

PUBLISHED BY
Ria University Press
PO Box 20170 at Stanford
Palo Alto, California 94309 USA

www.ria.edu/rup

Distributed by Ingram
Available from most bookstores and all Espresso Book Machines

---

Copyright © 2010 by Charles Tandy

Hardback/Hardcover ISBN-13: 978-1-934297-08-7

Paperback/Softcover ISBN-13: 978-1-934297-09-4

TO

two philosophers

Thomas O. Buford, Ph.D.
Jack Lee, Ph.D.

and all persons in search of philosophic understanding

Cartoon by Grea Korting <sangrea.net>. Reprinted by permission.

ABOUT THE AUTHOR

Dr. Charles Tandy received his Ph.D. in Philosophy of Education from the University of Missouri at Columbia (USA) before becoming a Visiting Scholar in the Philosophy Department at Stanford University (USA). Dr. Tandy is author or editor of numerous publications, including the ***Death And Anti-Death*** set of anthologies from Ria University Press. Dr. Tandy, along with Nobel Laureates and others, is a member of the Board of Advisors of the Lifeboat Foundation. Dr. Tandy is Senior Faculty Research Fellow in Bioethics at Fooyin University (Taiwan) where he serves on the Faculty of History and Philosophy and on the Medical Humanities Research Faculty. Dr. Tandy's CV is located at <http://www.ria.edu/cetandy/cv.html>. Also see: <www.segits.com>.

Contents

Book Summary   2

Dedication Page   5

About the Author   6

Acknowledgements   8

Forward   9

1. The End of the World as We Know It   15
2. Toward a New Theory of Personhood   21
3. N. F. Fedorov and the Common Task   45
4. Unburying the Dead   59
5. Is the Universe Immortal?   63
6. Earthlings Get Off Your Ass Now!   71
7. Ettinger's 1964 Thesis   85
8. My Dog Is a Very Good Dog   97
9. The Emulation Argument   115
10. Extraterrestrial Liberty and the Great Transmutation   131
11. A Time Travel Schema and Eight Types of Time Travel   147
12. Teleological Causes and the Possibilities of Personhood   163
13. Terrestrial Peoples, Extraterrestrial Persons   177
14. What Mary Knows   189

Index   205

ACKNOWLEDGEMENTS

The author, Dr. Charles Tandy, gratefully acknowledges support and assistance from the following:

- College of Humanities and Social Science, Fooyin University (Taiwan)

- Research Center for Medical Humanities, Fooyin University (Taiwan)

- R. Michael Perry, Ph.D., Society for Universal Immortalism (USA)

Foreword

21st Century Clues: Essays in Ethics, Ontology, and Time Travel consists of 14 of my papers published in the first decade of the 21st century (the years 2001-2008). The threads woven (the themes developed) become richer or more insightful, I believe, as they are woven, rewoven, and rewoven again over the eight-year period. This is why I use the term "clues" (rather than "speculations") in the title.

Abstracts for each of the 14 chapters follow:

Chapter 1
The End of the World as We Know It
Pages 15-20
The End Of The World As We Know It: Our Parental Responsibilities To Transhumanity. Apparently we are now in a time of metamorphosis from humanity to transhumanity, and many members of humanity living today will personally experience their own biomedical transformation from mortal being to transmortal being (indefinitely long life and health). This essay looks at the question of our worldwide parental responsibilities as we approach "the end of the world as we know it." Worldwide suicide is one possibility. In complex systems, the results of our actions can unpleasantly surprise us. We urgently need social science research on general systems, including specifically on the social system of planet earth.

Chapter 2
Toward a New Theory of Personhood
Pages 21-44
Toward A New Theory Of Personhood. This essay discusses various matters such as the objective ethical interests of persons, and the harm that death is -- based on a new theory of personhood developed herein. The Ultimate Person ("Up") is hypothetical but yet is a regulative idea: Following the will of Up is **not** hypothetical but a categorical imperative. We are to advance the objective interests of persons, including dead persons, by advancing science for the purpose of eventual resurrection of all dead persons. Via the advancement of our objective empirical interests, we learn to regulate nature; via the advancement of our objective ethical interests, we turn the world from indifference into love.

Chapter 3
N. F. Fedorov and the Common Task
Pages 45-58

N. F. Fedorov And The Common Task: A 21st Century Reexamination. First of all I make a few remarks about N. F. Fedorov and his philosophy of "the common task." After that I in effect reexamine his general project by looking toward future science and technology from our present perspective, the early 21st century. The nine basic sections to the article are entitled: 1. Fedorovian Philosophy; 2. Computers And Transhumans; 3. Feinberg's Law And World Catastrophe; 4. Expanding Beyond Our Cradle Earth; 5. Molecular Nanotechnology; 6. Suspended Animation And Transmortality; 7. Scientific Universal Resurrection; 8. Time Travel; and, 9. Prediction And Retrodiction.

Chapter 4
Unburying the Dead
Pages 59-62

Unburying The Dead: Posthumous Harms And Posthumous Benefits -- A Solution To The Missing Subject Problem. When is one harmed -- and who is the subject of the harm -- if one's harm is posthumous? The past is a fixed unity that will always exist even if the universe dies. Thus a dead person may be characterized as presently existing fact-information. Accordingly, there is no "missing subject" problem. Following N. F. Fedorov, the living should love all dead persons and take into account the real interests of dead persons. Resurrecting all dead persons by future scientific means would benefit the dead.

Chapter 5
Is the Universe Immortal?
Pages 63-70

Is The Universe Immortal?: Is Cosmic Evolution Never-Ending? This five-part paper describes Eric J. Chaisson's interdisciplinary account of the rise of complexity in nature, *Cosmic Evolution*. Part one asks, "Is the universe a deterministic machine?" explaining Chaisson's negative answer. Part two, "Is cosmic evolution a good personal philosophy to live by?" takes issue with Chaisson's positive response. Part three, entitled "three thought experiments -- and their surprising results when scientifically tested" recounts a bit of history not covered by Chaisson. Part four explains Chaisson's formulation of "free energy rate density" as a first step toward characterizing the engine or fuel of cosmic evolution. Part five explains why Chaisson believes a never-ending perpetually-dynamic universe should not be ruled out.

Chapter 6
Earthlings Get Off Your Ass Now!
Pages 71-84
    Earthlings Get Off Your Ass Now!: Becoming Person, Learning Community. This paper is about the present age and how to prevent doomsday. Part One is entitled "an age of absurdity and uncertainty." The present age as absurd and uncertain is suggested by historical events specified herein. Part Two is entitled "an age of anti-absurdity and uncertainty." Certainty is absurd; uncertainty is anti-absurd. It is not wise to put all of humanity's eggs (futures) into one basket (biosphere). Construction of large, comfortable, permanent, self-sufficient Extra-terrestrial Green-habitat Communities (not to be confused with space stations) is feasible this century. Ratification of a proposed Space Treaty now, before we weaponize space, would prevent an arms race in outer space, our future home.

Chapter 7
Ettinger's 1964 Thesis
Pages 85-96
    Ettinger's 1964 Thesis: Indefinitely Extended And Enhanced Life (Immortality) Is Probably Already Here Via Experimental Long-Term Suspended Animation. In 1964, Robert Ettinger argued that immortality (as indefinitely extended and enhanced life) is probably already here via cryonic hibernation (experimental long-term suspended animation). This paper performs three functions. First, it introduces and summarizes Ettinger's *The Prospect of Immortality* (1964) for readers of the 2005 (reprinted) edition and others. Secondly, it presents and paraphrases Ettinger's arguments using current 21st century terminology. The paper also contains a new bibliography on cryonics.

Chapter 8
My Dog Is a Very Good Dog
Pages 97-114
    My Dog Is A Very Good Dog – Or – The Unprecedented Urgency Of New Research Priorities To Dismantle Doomsday And Cultivate Transhumanity. Particular cultural traditions have informed each civilization's felt educational needs to become "us" or "human" (instead of barbarian) or to become "educated" or "transhuman" (instead of merely human). The twentieth century surprised many of us with its world wars and doomsday weapons (WMDs). If we survive all doomsday dangers over the next few years and decades and centuries, then our future as humans or transhumans may be longer – much longer – than the mere 10,000 years of past civilizational existence. Our pasts

are short and almost non-existent compared to the potential reality of a very long future. This paper explores the educational implications of such a complex reality.

Chapter 9
The Emulation Argument
Pages 115-130

The Emulation Argument: A Modification Of Bostrom's Simulation Argument. Bostrom's posited computer simulated world would be real and the simulated persons in it would have mentality. However Searle has shown that computers (i.e. mere algorithmic symbol manipulators) with mentality are impossible. This paper also shows that Bostrom's conclusion is unreliable regardless of whether Searle is right or wrong. A modified or alternative argument is then presented. The emulation argument postulates possible entities not unlike Bostrom's supposed computer simulations with mentality. Since mere computers alone cannot directly produce the supposed simulations, other devices (machines other than mere algorithmic symbol manipulators) may be able to produce said emulations. The paper also presents speculative discourse on related metaphysical and moral issues.

Chapter 10
Extraterrestrial Liberty and the Great Transmutation
Pages 131-146

Extraterrestrial Liberty And The Great Transmutation. Doomsday weapons exist: Since 1945 we have been living in the era of the great transmutation. Perhaps humanity will give birth to transhuman superintelligence sooner rather than later. How may humanity be a good parent to its transmortal offspring? Is the surface of a planet the ideal place for transcivilization? What might the political structure of a "realistic utopia" be like? The author answers these three questions and invents a transhuman political philosophy, some of which can be feasibly implemented now. The author concludes that if indeed humans do their part now, then the great transmutation is doable -- a transhuman world of liberty at stable peace is possible.

Chapter 11
A Time Travel Schema and Eight Types of Time Travel
Pages 147-162

A Time Travel Schema And Eight Types Of Time Travel. Based on specified logical, ontological, and other relevant considerations, it is concluded that in the very-long-run: (1) forward-directed time travel capacity is highly likely; and, (2) past-directed time travel capacity is likely. Four logically possible forward-directed, and four logically possible past-directed, types of (hypothetical) time machines are identified. Two different approaches (the "practical"; the "bi-temporal") are utilized in attempting to characterize the meaning of time travel. It apparently turns out that the concept of "embedded-subjective time" (i.e. the embedded-temporality of the human time-traveler, as distinguished from either merely-subjective time or literal-wristwatch time) is especially helpful in characterizing whether time travel did or did not occur in a particular circumstance.

Chapter 12
Teleological Causes and the Possibilities of Personhood
Pages 163-176

Teleological Causes And The Possibilities Of Personhood. In part one ("The Many-Multiverses (M-M) Model"), I postulate certain metaphysical and existential assumptions which serve to model the larger world (or multiverse of multiverses) in which we are said to be embedded. In part two ("The Possibilities Of Personhood"), I describe the general sorts of individual physical (time-space) entities that exist in this region of this universe or that may exist in other or future regions of the multiverse. (An anti-speciesist stance results.) Finally, in part three ("Super-Persons As Teleological Causes"), I take for granted the context outlined in parts one and two so as to consider ethical-teleological issues related to the emergence of super-persons (quasigods) from persons.

Chapter 13
Terrestrial Peoples, Extraterrestrial Persons
Pages 177-188

Terrestrial Peoples, Extraterrestrial Persons. The political structure of Earth, which is neither a Law of Peoples nor a Law of Persons, is unworkable. But at this unique point in history it is both desirable and feasible to establish a Terrestrial Law of Peoples along the lines of Kant; Rawls; and, Daalder and Lindsay. The political structure of Space, which is neither a Law of Peoples nor a Law of Persons, is unworkable. But at this unique point in history it is both desirable and feasible to establish an Extraterrestrial Law of Persons along the lines indicated herein.

Chapter 14
What Mary Knows
Pages 189-204

What Mary Knows: Actual Mentality, Possible Paradigms, Imperative Tasks. In part one (of two parts) I show that any purely physical-scientific account of reality must be deficient. Instead, I present a general-ontological framework that should prove fruitful when discussing or resolving philosophic controversies; indeed, I show that the paradigm readily resolves the controversy "Why is there something rather than nothing?" In part two, now informed by the previously established general ontology, I explore the issue of immortality. The analysis leads me to make this claim: Entropy is a fake. Apparently the physical-scientific resurrection of all dead persons is our ethically-required common-task. Suspended-animation, superfast-rocketry, and seg-communities (Self-sufficient Extra-terrestrial Green-habitat communities, or O'Neill communities) are identified as important first steps.

The book's **Index** begins on page 205.

Chapter 1
The End of the World as We Know It

"The End Of The World As We Know It: Our Parental Responsibilities To Transhumanity" was first published in 2001 and is here reprinted by permission.

As this **Guide** {*Doctor Tandy's First Guide To Life Extension And Transhumanity*} was about ready to go to press in September 2001, World War Three began with a coordinated set of attacks on the World Trade Center in New York City. Cultural historians tell us that "the twentieth century" began with World War One (1914). Thus it seems that "the twentyfirst century" -- indeed, "the third millennium" -- began with World War Three (2001). The immense differences between World War One (WW1) and World War Three (WW3) suggest that our metamorphosis from humanity to transhumanity is neither guaranteed nor without difficulties.

Teleological or eschatological issues -- whether approached from a traditional paradigm (religious or scientific) or from some new paradigm now being invented/ discovered -- continue to play major roles in the story of human history. This book {*Doctor Tandy's First Guide To Life Extension And Transhumanity*} discusses such issues -- but no doubt many of our readers will come away believing (if they did not before) that eschatology and the far future are complex subjects about which we know little with certainty. That can be a healthy attitude -- and, so far as I am concerned, faith (as distinguished from certainty) in a traditional religious or scientific paradigm or in some new paradigm can be healthy too.

Not without reason, cultural historians sometimes speak of an "imaginative" or "superstitious" Medieval period of Western civilization filled with contradictions (such as love and violence). Indeed, with their certainties of religious belief came many barbarous or terrorist attacks on non-Christians and fellow Christians. (Similar barbarous behavior from those with anti-religious certainties have also been experienced in human history.) On the other hand, we sometimes forget that many of the Medieval folks thought it their parental or custodial responsibility (their God-given duty) to improve humanity, perhaps even transform it into utopia. Indeed, many of the greatest scientific discoveries of the humans were due in part to the scientists' deep faith in their God or religion (at times despite attacks from those who supposedly believed in a similar God or religion).

In this time of transition from civilization to transcivilization, what are our "God-given duties" or "parental responsibilities" toward the present and future (and past)? According to one scenario that I find easy to believe, the transition is really a metamorphosis from humanity to transhumanity, and many members of humanity living today will personally experience their own biomedical transformation from mortal being to transmortal being (indefinitely long life and health). What are our worldwide parental responsibilities as we approach "the end of the world as we know it"? Let me share some of my thoughts:

One

I am not the only one who has pointed out repeatedly over a period of decades that worldwide suicide is possible. I have said that in addition to the possibility of nuclear war between nation-states, a nuclear or biological or chemical or other attack by a group of terrorists or perhaps even by a single terrorist might result in worldwide suicide (the extinction of all life or all "intelligent" life we know of). We need to better articulate, and act on, our responsibilities in this regard.

Two

Catastrophe, then, is one possibility. But barring catastrophe, and assuming a plausible scenario of a successful twentyfirst century metamorphosis, many humans living today will personally experience their own transition from human to transhuman. We need to better articulate, and act on, our responsibilities related to successful transition.

Three

Our responsibilities toward transhumanity in the traditional human parenting sense are very limited. For one thing, we are both parent and child. Many humans living today (perhaps me; perhaps you) will be transformed and transformed again. The transhuman child will be superintelligent compared to the human parent. (Indeed, most humans would have chosen to be transformed into transhumans.)

Traditional human parenting includes protecting the child and imparting proper values:

- Protecting The Child From Death And Danger. This would still seem to make sense, except that the transhuman "child" would know more about this than the human. Thus if there is a conflict of opinion, the transhuman child, not the human parent, would probably know best.
- Imparting Proper Values To The Child. Again, at first, this seems to make some sense. But in fact the child would already know more about philosophy, science, religion, and values than the parent.

We are both parent and child as we transform from human to transhuman -- thus our responsibilities are to a successful transformation and re-transformation of self, others, and society. This requires, I think, a radical openness to learning (the superintelligent child always reinventing itself). So it seems that many of our responsibilities go directly to doing what we can to make the twentyfirst century transition period result in success. With assistance, perhaps, from Kenneth Boulding's *The Meaning of the Twentieth Century*, we can begin by attempting to identify the barriers to metamorphosis.

Four

Sometimes we do something, expecting one result, only to get something very different. Perhaps the terrorists who began WW3 thought the prideful power-hungry United States or Western World or Human Species would begin to disintegrate from within. But in fact the result was very different. Perhaps some Western strategists thought that the Western World should unite to defeat the terrorists. Such strategists might have thought they were being unusually intelligent by seeking a coalition rather than having the Americans go it alone. But in fact even this supposedly more enlightened strategy might easily backfire. Non-Westerners may ask: If the West is already the master of the world, why does it want to also now police the world instead of ending poverty, ignorance, and economic underdevelopment? Have not the Westerners already starved or killed too many Non-Western men, women, and children? So the Western strategist thinks again: Maybe we need a Western "Marshall Plan" to help the Middle East? But one may then ask: Why the **Middle East** -- and why a **Western** "Marshall Plan"? Combine questions like these along with the possibility of worldwide suicide, and one can see why WW3 must be the most complex war in human history.

Let me suggest that WW3 become a long-term series of campaigns (as stated by U. S. President George W. Bush), but that its goal be an international effort to make a successful transition from a world of terror to a world of stability. As we look at history, we typically look at it with rose-colored, or at least tinted, glasses. We judge our own group by its good intentions and we judge another group by its cruel behavior. When we take off our glasses, we find to our surprise that **differential development** explains more of the history of humans than we had supposed. (See, for example, Jared Diamond's ***Guns, Germs and Steel***.) Where are the underdeveloped areas today that need attention? How exactly do the twin issues of differential development and of the advancement of science-technology fit together as we strive toward a stable world of transhumans? When we think of differential development, we need not always think in terms of developed or underdeveloped **countries**. For example, there are developed areas in the Middle East and there are underdeveloped areas in the United States. As a practical matter, we need to build an infrastructure (in each underdeveloped area or for every person worldwide?) that would connect each individual's pursuit of happiness with positive (or at least relatively non-negative) results for the society in which they live. In history past, we find that there were underdeveloped areas that became developed, and that there were developed areas that became underdeveloped.

I am of the opinion that in the present moment we need to strengthen the **family** unit: Make it so parents **can afford** to spend time with their children. Human children (and possibly transhuman children?) grow up with a huge disadvantage if they too rarely experience parental love. Speaking for myself, I want children who are superintelligent in a wide variety of ways, both intellectually (IQ) and emotionally (EQ). Low EQ children or children who have never or rarely experienced parental love may be foredoomed to knowing what love is; they may be forever mistaken as to what love is. Some of them will seek to metamorphose from human to transhuman; we may need to give them special love and attention. The results may be uncertain (some high IQ, low EQ transhumans?); some of the results, human or transhuman, may be unloving or anti-loving intelligence and barbarous, cruel behavior. (What is love? For one perspective, see Soren Kierkegaard's ***Works of Love***.)

Now back to the topic of mistaken results. Good motives are not good enough for good results. We need social science research on general systems and, more specifically, on the social system of planet earth. Without such intelligence, we (as individuals, groups, and nations) will too often be unpleasantly surprised by the results produced by our actions.

Five

Relatively few financial records were lost in the World Trade Center attack. Too many human persons were lost. Copies of the financial records were stored in known alternative locations. This was not true of the human persons; they existed in biological but not digital form.

At this point in time, we do not know (at least I do not know) if, either sooner or later, it will become possible to "digitalize" persons. But I do believe I know, on the one hand, that cryonic hibernation facilities have existed for decades; and I believe I know, on the other hand, that the dinosaurs became extinct because they did not have a space program sufficient to prevent their extinction. The right kind of human space program -- using technology in our grasp for decades -- would insure a human flourishing and expansion in the universe regardless of what happens to planet earth. (For example, see the late Dr. O'Neill's *The High Frontier: Human Colonies in Space*.)

Six

Should transhumanity be our goal? Technological speed can kill -- but it can also save lives. For the sake of the present discussion, let me make a distinction between transhumanity and transmortality. We, for example, identify the term **transhumanity** with wanting to know everything about everything. We don't know what a robust transhumanity would be like: It is probably not merely beyond our comprehension, but beyond our ability to comprehend until such time as we become transhuman. But relatively early on -- before transhumanity has had a chance to mature, perhaps while we are still more or less in the transition stage and still identify ourselves as human -- we will probably achieve **transmortality** (the cure and conquering of all disease, including the disease of aging to death -- that is, indefinitely long life and youthful health).

At that point it may be appropriate for us to consolidate our gains, make sure we have a **stable** world in which **every** person is (if this is their desire) healthy and wealthy (and wise?). A stable world of transmortals would seem to me to be achievable before the century is out. Once achieved, it would be an occasion for us to congratulate ourselves. Then we can choose the rate or metabolism we consider appropriate for achieving a robust transhumanity or series of metamorphoses. At least from my present vantage point, I am convinced that this new stable world of transmortals should set for themselves the common task of resurrecting all dead persons who have ever lived and offering them transmortality.

My "crystal ball" (such as it is) does not go much beyond this (although some of the contributions to the present volume {*Doctor Tandy's First Guide To Life Extension And Transhumanity*} have started me thinking). It does occur to me, however, that it may be more difficult for humans (or transmortal quasi-humans) to achieve a stable world than for transhumans. If this is so, then to be very successful in achieving a stable world of transmortals, we would want a robust transhumanity sooner rather than later. It's also possible that these considerations are presently irrelevant or will become meaningless in that all of these matters may take on a life of their own. For example, at least presently (and at least for decades) we are not in control of technology: Technological advance is in control of us. Barring catastrophe, we are perhaps talking about slowing down or speeding up things a bit. But in any case (barring catastrophe) it's difficult for me to see how transmortality could not result before our century is out. Twentieth century biomedicine extended the **average** lifespan. Twentyfirst century biomedicine will extend the **maximum** lifespan; this unprecedented extension of life and health, as preface to transmortality and transhumanity, may well arrive in the early decades, not the latter decades, of the century.

Chapter 2
Toward a New Theory of Personhood

> "Toward A New Theory Of Personhood" was first published in 2002 and is here reprinted by permission.

Free agency or moral agency is arguably the essence of personhood. I say "arguably" because the concept of "person" or "personhood" (compared to some other concepts) is highly contested by today's philosophers. For example, are dogs and cats (and humans) "persons" even if they lack moral or free agency? It is **not** my intent here to formulate a defensible or definitive **definition** of personhood. Nevertheless it **is** here my intent to work toward formulating a new **theory** of personhood based in part on Jack Li's new theory of harm to persons.[1] In the process we may learn more about the meaning and implications of personhood even as we continue to offer no persuasive or convincing definition of the concept.

Jack Li: The Objective Interests of Persons

Jack Li {Jack Lee} has explained the concept of **objective interests** as follows:[2] "Y is in X's [objective] interests." "It is possible for someone to be interested in something that is not really in his **objective** interests. It is also possible for something to be in his **objective** interests regardless of the fact that he is not presently interested in it."

What is in the objective interests of persons? An obvious answer is that it is in the objective interests of persons to advance toward ultimate personhood. This includes the advancement of empirical-scientific learning and of ethical-moral learning.

Ancient Knowledge and the Advancement of Learning

Twentieth and twenty-first century science-technology suggests that matters are indeed subject to radical improvement if present and future persons act wisely -- and that matters are indeed subject to radical catastrophe if present or future persons act unwisely. Until recent centuries, it was common among many of the peoples of the planet to assume that the structure of the world and one's place in it were more or less fixed or static. Or that major changes were beyond the immediate control of mere humans. Nature, God, or gods were very powerful compared with humans; presumably it should be that way. The **learning of ancient knowledge**, and handing it down to future generations, was the way of wisdom. But more recently, with the advent of modern science-technology, we have instead viewed the **advancement of learning** as the way of wisdom.

Thus today we are generally inclined to think that self improvement, world betterment, and the advancement of learning are both possible and in our objective interests. But although human persons seek their objective interests, they also engage in self-deception. Whether because of self-deception, weakness of will, ignorance, or some other reason, we may not always know what our objective interests are or how to pursue them; if we do know how, we may nevertheless not pursue them.

Should our emphasis be on **preventing** the impairment of our objective interests **OR** on **pursuing** our objective interests? Pursuing our objective interests is the answer, for it subsumes or includes preventing the impairment of our objective interests. The practical situation can sometimes require as our best choice being harmed or risking harm in the pursuit of the advancement of our objective interests. (In another situation, our best choice may be not to risk harm or not to be harmed.)

Persons-In-Relationship

If the advancement of learning is in the general objective interest of persons, what are its implications for personhood and living a life worth living? Human persons interact with each other, thereby creating organizations (families, churches, universities, governments, etc.), including roles appropriate to those organizations.[3] Human persons interact with each other, thereby learning what it means to be persons, hopefully always advancing in their learning of personhood.

It is not reasonable to expect each human person to be an expert in every field and a master of every role. Moreover, each person-to-person interaction is more or less unique. Yet for a society and the persons in it to survive and flourish, there is a sense in which there must be a coordination or balance of learnings. For example, lack of advance in the learning of matters ethical can prevent advance in the learning of matters empirical (science-technology). Indeed, there are still some people who believe that drought, earthquake, crop failure, physical drudgery, AIDS, cancer, age-related disability, and death are the will of God. On the other hand, advance in the learning of matters empirical can result in the extinction of all life or all intelligent life on the planet -- if there is no corresponding advance in the learning of matters ethical.[4] Indeed, there are still some people who believe that the value of scientific progress makes other values irrelevant.

Human persons do not advance in their personhood except in the proper environment. Although a human person cannot be an expert researcher in all possible fields, a society of human persons can achieve a reasonable balance of expert researchers. Although a human person cannot be a master role-model for all possible roles, a society of human persons can achieve a reasonable balance of master role-models. A well-balanced society can help provide the proper environment for each person to survive and flourish. Indeed, if you had lived in a social environment of 10,000 years ago, you would be a very different person. If Einstein had lived in a different time or place, he might not have survived and flourished. Thus there is a sense in which all human persons are necessarily social; this is a way individuals advance in their personhood. Accordingly, as we consider the objective interests of persons and the specific objective interest "the advancement of learning," we must see human persons as individuals-in-relationship and as dependent on a proper environment. Like other animals, a human person will not flourish in a dismal physical environment. Likewise, a human person will not flourish in a dismal social environment and may not be able to advance in the learning of personhood. Human persons are neither individual atoms nor social insects, but are individuals **in relationship** to other individual persons. A specific person at a specific time-place exists in a specific environment of relationships.

From these considerations it follows that it is in the objective interest of human persons to live in a society (social environment) that promotes the advancement of learning at both the societal and individual levels. (Not only may individuals advance in their learning or development, but also societies may advance in their learning or evolution.) Moreover, a balanced society of human persons will promote the advancement of both empirical learning and ethical learning (rather than one without the other). At this point in time in human history (the early twenty-first century), part of the advancement of learning must specifically address itself to possible self-extinction of all intelligent life on our planet and to the flourishing of every human person. This requires redirection and intensification of the advancement of learning in matters both empirical and ethical. A restructuring of our ethics will guide us in a restructuring of our empirics.

The Objective Ethical Interests of Persons

Although the present study will not investigate directly issues of political philosophy, our insights may not be altogether irrelevant to some such issues. Above we found that human persons are "individuals-in-relationship." Questions arise as to how such persons should live their lives and treat each other. What are the objective interests of human persons in terms of ethical advancement and good decision-making?

For **empirical** advancement, many humans turn to what is loosely called "the Scientific Method." Sometimes it seems each practitioner of "the Scientific Method" (or "Right-Good Empirics") has a unique view as to exactly how this "method" (or set of methods or right-good empirics) should best be formulated. For **ethical** advancement, many humans turn to what is loosely called "the Golden Rule." Sometimes it seems each practitioner of "the Golden Rule" (or "Right-Good Ethics") has a unique view as to exactly how this "rule" (or set of rules or right-good ethics) should best be formulated.

It may be in our objective interests to hold and use a variety of different formulations of empirical methods or approaches. Likewise, it may be in our objective interests to hold and use a variety of different formulations of ethical rules or approaches. Of course, in a particular situation we will have to make one decision rather than another. By being familiar with the various formulations (empirical or ethical) and how to use them, our decision may be more difficult (sometimes) yet more likely to be in our objective interests.

Some formulations of the Golden Rule (or Right-Good Ethics) give reference to Divinity, some do not. Others believe that although ethical rules or decision-situations are indeed related to Divinity, they can nevertheless be formulated without such reference; for example, one may or may not include a reference to Divinity in one's "do" or "do-not" rule or rules. By way of introduction to my unique formulation of the Golden Rule or Right-Good Ethics (based on the objective interests of persons), I refer you to a dialogue between Bertrand Russell and F. C. Copleston:

- RUSSELL: Obviously the character of a young man may be -- and often is -- immensely affected for good by reading about some great man in history, and it may happen that the great man is a myth and doesn't exist, but the boy is just as much affected for good as if he did.[5]

- COPLESTON: In one sense he's loving a phantom that's perfectly true, in the sense, I mean, that he's loving X or Y who doesn't exist. But at the same time, it is not, I think, the phantom as such that the young man loves; he perceives a real value, an idea which he recognizes as objectively valid, and that's what excites his love.[6]

Human persons are individuals-in-relationship and dependent on a proper environment to survive and flourish. Self-improvement and world-betterment are obviously in the objective interests of persons. It is in the objective interests of persons to advance toward **ultimate personhood** and **ultimate cosmos-hood**. Regardless of whether something like the **Ultimate Person** has, does, or will exist, the Ultimate Person nevertheless serves as a golden guide or regulative idea to advance our objective interests. Likewise, regardless of whether something like the **Ultimate State-of-affairs** has, does, or will exist, the Ultimate State-of-affairs nevertheless serves as a golden guide or regulative idea to advance our objective interests.

Consider the properties (characteristics or traits) of the Ultimate Person ("**Up**"). Consider the properties (characteristics or traits) of the Ultimate State-of-affairs ("**Us**"). Consider how self and world are to advance from here to there (**How to Advance**: "**Ha**"); Up (the Ultimate Person) informs the proper means to advance our objective interests.

A further word about terminology. "**Gohs**" (plural) are the General (or prima facie) Objective interests of Human or finite persons, and a "**Goh**" (singular) is a General (or prima facie) Objective interest of Human or finite persons. Flagrant (or prima facie) Objective Harm(s) to human or finite persons are "**Fohs**" (plural) or "**Foh**" (singular). In accordance with what we have said previously about reasonable priorities, Fohs (prima facie impairments of objective interests) are a kind of negative subset of Gohs (prima facie objective interests). Thus Gohs (General Objective-interests of Human-persons) include or subsume Fohs (Flagrant Objective Harms).

My explicated-and-articulated list of Gohs is always open to improvement (subtraction; addition; modification). This openness to improvement not only applies to my list of Gohs (the General Objective-interests of Human or finite persons) but also to my conceptions of Up (the Ultimate Person), Us (the Ultimate State-of-affairs), and Ha (How to Advance). Our explicated-and-articulated lists regarding Gohs, Up, Us, and Ha, are prima facie but are not arbitrary. We have help from those who have gone before; we continue the great "never-ending" conversation. Moreover, the four lists (Gohs, Up, Us, and Ha) interact so

as to produce more coherent formulations. More pointedly, theologians and philosophers have engaged in dialogue over the centuries about the character traits of human beings, ethical persons, and Divinity. Thus informed, I offer an initial list of prima facie properties of the Ultimate Person (Up):
1. Tremendous Creative-energy
2. Tremendous Self-control
3. Tremendous Knowledge
4. Tremendous Power
5. Tremendous Wisdom
6. Tremendous Love-goodness
7. Tremendous X-unknown

To put it another way, it is in the (prima facie) objective interests of finite persons to increase and enhance their: 1) Creative-energy; 2) Self-control; 3) Knowledge; 4) Power; 5) Wisdom; 6) Love-goodness; and, 7) X-unknown. (**X-unknown**: The six previous properties or characteristics were not meant to be exhaustive; beyond this, much is unknown or unknowable to mere human or finite persons.)

We need a balanced approach of becoming both more powerful and more good, as distinguished from pursuing one without the other. If the "center" of our self and of our world is to "hold," then a coherent or balanced approach is required. This would be, so to speak, the voice of Up's Wisdom speaking up. Does anyone doubt that our "Up" working hypothesis, our just-stated conception of the Ultimate Person, makes some sense as a rough or pre-penultimate approximation? That is to say, does it make any sense to say that the Ultimate Person would **not** have tremendous creative-energy, would **not** have tremendous self-control, would **not** have tremendous knowledge, would **not** have tremendous power, would **not** have tremendous wisdom, would **not** have tremendous love-goodness, would **not** have tremendous x-unknown? Without these qualities, would we refer to this person as "ultimate"?

What are the implications of the Up hypothesis for personal relationships and for interactions among human persons? [7] Up's character trait "love-goodness" immediately comes to mind. Being loved and advancing in one's love-goodness is certainly an important aspect of learning to be a person, an individual-in-relationship. In terms of logic, Up first loved us. Obviously Up's love-goodness is ultimate: Up's love of persons is unconditional, creative-active, and universal. By opening ourselves up to Up's love and actively-creatively experimenting with this love in our everyday lives, we can advance in our learning of true love. We can learn to love our enemies. We can try to help them learn of their objective interests. We can try to pursue our objective interests; this includes trying to help all persons pursue the objective interests of every person. The statement "love your neighbor as yourself" is one articulation of this imperative.

Up is a golden guide or regulative idea. **Up is hypothetical** in that we do not argue Up's existence or non-existence. But the nature of the Up idea is such that it is not merely another idea among others. Indeed, **following the will of Up is not hypothetical**: It is an unconditional or categorical imperative.

Disobeying the will of Up means one has done wrong rather than right; one has failed to act in the objective interests of persons. To be sure, we may not know the will of Up in a given situation. Or we may "know" the will of Up only later to decide that our "knowledge" was faulty. We must always be open to the will of Up and to changing our conceptions of Up and of Up's will. Indeed, we have previously identified "x-unknown" as one of Up's properties.

We are unconditionally commanded or guided to **feel**, **think**, **will**, and **act** lovingly. Two people (say, Robin Hood and Crime Boss) may act in the same way or do the same thing (rob banks). But we interpret their actions differently if we believe one acted out of love and the other out of ethical rebellion or vileness or lack of love. More fundamentally, what does or does not gain our **attention** to begin with, including our **interpretation** of it, can be dependent on our love or lack of love.

Love is creative-active and is expressed in works of love. Some may think they know about ethical matters and declare a work to be unloving, unethical. But true love, being active-creative, follows the expectations of neither saints nor sinners. True love attempts to creatively-actively understand and creatively-actively follow the will of Up. It places the will of Up above the habits and expectations of saints and sinners alike. One can choose either to act in the objective interests of persons or to rebel against acting ethically. We must learn to advance in our personhood and to redirect our **attention, interpretation, feeling, thinking, willing, and acting** -- away from merely subjective interests to creative-active love and objective interests.

Human persons are able to consciously engage in ethical contemplation or moral reflection about alternative courses of action. Such reflection, if sufficiently practiced and developed, can prepare one for more advanced reflection or for tentative action. We are commanded by Up to engage in such difficult ethical reflection and decision-making.

The command is a fact or non-hypothetical imperative even if Up is hypothetical. Up does not essentially or necessarily tell us to ask for forgiveness but rather commands us to learn to live actively-creatively-lovingly together in our objective interests.

Love always remains new or active-creative. The advancement of living, learning, and loving is not about self or others but about self and others. Self-improvement is not a resting point but a beginning for greater self-improvement. World-betterment is not a resting point but a beginning for greater world-betterment. Our findings or conclusions are tentative rather than absolutely conclusive. Whether due to absolute conclusions or lack of absolute conclusions, one may be tempted to despair. Yet love never gives up. Love says "Do not despair; **now** (the present situation) is always the right time for love." But the right-good decision or loving action may well surprise those who think they know with absolute certainty what love is.

We impair the advancement of our objective interests when we impair our objective growth. One who does not love, or no longer loves, has done great (objective) harm to oneself. One "can be deceived in believing what is untrue, but on the other hand, one is also deceived in not believing what is true."[8] "To cheat oneself out of love is the most terrible deception; it is an eternal loss for which there is no reparation."[9] Love is always new and never gives up: You **harm yourself** if you are not lovingly trying to improve yourself and the world.

Our findings or clarifications above include the following: There is a sense in which self-improvement and world-betterment are in the objective interests of finite persons. The regulative idea of the Ultimate Person ("Up") can be used to help explore the implications of self-improvement and world-betterment. The regulative idea of the Ultimate State-of-affairs ("Us") can be used to help explore the implications of world-betterment and self-improvement. The idea of the General Objective interests of Human or finite persons ("Gohs"), of the Ultimate Person ("Up"), and of the Ultimate State-of-affairs ("Us") -- along with the notion of How to Advance ("Ha") individually and collectively toward Gohs, Up, and Us -- are obviously conducive to the objective actualization of personhood. **We now report a fifth approach**: Call it "**Card**" for Coherence Analysis, critical Reflection, and never-ending Dialogue. We can interactively compare the four explicated-and-articulated lists (Gohs, Up, Us, and Ha) so as to produce more coherent formulations -- yet such coherence cannot be the ultimate word. Coherence may be a step toward the advancement of personhood and in the objective interests of persons. It may even give the appearance of ultimate truth or objective reality. But coherence is coherence, not ultimate reality or objective truth. Thus we must always be on the lookout for more advanced explorations and alternative formulations of objective reality and our objective interests. Accordingly, we do not become lazy, careless, or arrogant -- rather, with our perhaps tremendous strides toward self-improvement and world-betterment, we become more and more humble. Critical reflection and never-ending dialogue are not optional; indeed, perhaps they are at the heart of personhood and the advancement of our objective interests.

Above we identified prima facie traits or characteristics of the Ultimate Person (the "Up" regulative idea): Tremendous creative-energy; self-control; knowledge; power; wisdom; love-goodness; and, x-unknown. Further, we explored Up's characteristic "love-goodness" (with a little help from "creative-energy" and "x-unknown"). Up is **hypothetical** in the sense that we do not argue Up's existence or non-existence. Yet, for human or finite persons, **following the will of the Ultimate Person is not hypothetical**: The will of Up is an unconditional or categorical imperative. Up is a regulative idea -- but in terms of logic, Up first loved us. The Ultimate Person's love of persons is obviously unconditional, creative-active, and universal (otherwise "Up" would not be "Ultimate"). Thus Up teaches, guides, directs us to actively-creatively love our enemies. The statement "love your neighbor as yourself" is one articulation of this imperative. "X-unknown" is one of Up's characteristics from the point of view of human or finite persons: Accordingly, we must always be creatively-actively open to the will of Up -- and to changing our conceptions of Up and of Up's will. A living human person can choose either to act in the objective interests of persons or to rebel against Up's will. Up commands us to live actively-creatively-lovingly together in our objective interests. One who does not love does great (objective) harm to oneself. Below we will find that the newly restructured ethics just outlined is helpful in the restructuring of our empirics.

The Advancement of the Objective Interests of Persons

There are two general categories of ways a person may be helped or harmed.[10]   A person may be helped or harmed by **persons**. And a person may be helped or harmed by **nature**.

Persons. One may be helped by persons. For example, a good parent or a good teacher may help you advance toward ultimate personhood. One may be harmed by persons. For example, a bad parent or a slanderous neighbor may harm your advancement toward ultimate personhood.

Nature. One may be helped by nature. For example, you develop biologically from infant to adult. One may be harmed by nature. For example, you develop age-related disabilities or become diseased.

If one is harmed, the category of harm (either persons or nature) indicates the kind of solution needed as to advance toward ultimate personhood. Unloving or bad relations among persons must be changed into loving or good relations among persons. And nature must be regulated so as to make it less harmful and more beneficial toward the advancement of personhood.

The advancement of the objective interests of persons requires assessing our present situation and flexibly planning reasonable steps to advance toward ultimate personhood. This means having a dream or life-plan. As one advances in one's life-plan, the dream itself will change or advance. As a child I dreamed of becoming a firefighter because I liked fast red vehicles with loud sirens. Later my dream was to secure a job for economic survival. Later still I decided that a meaningful career rather than mere economic survival was a good thing. Based in part on the work of Gisela Striker, it is obvious to Jack Li that generally "the fulfillment of our (well-considered) life plans or projects is in our [objective] interests. For these to be fulfilled (i.e. to live a **complete** life), we must have a sufficient life span."[11]

The newly restructured ethics outlined above can help inform our dreams or life-plans. It is in my objective interests to advance toward ultimate personhood. If it is in my objective interests to advance toward ultimate personhood, then a mere century of living is a grossly insufficient life span for a "complete" life. Given this context, it is obvious (prima facie or in general) that events like drought, earthquake, crop failure, smallpox, AIDS, cancer, age-related disability, and death are not in the objective interests of persons. As Jeff McMahan points out: "This is not the best of all possible worlds. ... Were it not for these various evils, each of us would enjoy the prospect of an indefinitely extensive succession of possible goods ... "[12]

Developing a life-plan is not exactly a novel idea, but it is good practical advice in our objective interests. Yet there is something new about a life-plan -- a life-plan in the objective interests of persons. Previously we took a mere one-century life-span for granted. It was said that via a well-formed dream or life-plan, "**one ought to choose a** coherent life in which the widest possible range of values is realized in accordance with a life-span" of one-century.[13] But today it seems clear we are historically near a great inflexion point that will take us far beyond previous human life and lifespan limitations -- that is, barring another real possibility, radical or worldwide catastrophe.

Death and Personhood

Epicurus (341-270 BC) argued that "death is nothing to us" because one must be alive in order to experience harm or be harmed.[14] But being dead means one is not alive. Thus, according to Epicurus, death cannot be a harm to the person who dies.

Below I will show that a person can be harmed in the absence of experiencing harm. I will also show that a person can be harmed after death. "**Harm**," according to Jack Li, "**is the impairment of objective interest.**"[15]

Can a person be harmed without experiencing harm? Indeed there are situations we can think of in which a person is harmed in the absence of experiencing harm. Consider the following cases:

(CASE: Enjoys Abuse -- by Jack Li)[16]
- Someone P is fed a drug by Q when he is sleeping. The process does not cause P any discomfort. After that, P is taken away to Q's place. Q treats P as a slave and often abuses him. Because the drug makes P enjoy what Q does to him, P likes to be Q's slave.

(CASE: Malicious Lies -- by Jack Li)[17]
- Suppose a person P had a lovely family -- with a beautiful and sweet wife, a clever and cute son, and a friendly dog. P was a good man and had a very good reputation which he was very proud of. Q was P's good friend. Two years ago, P went to an island to do some business for six months. After he left for this island, Q started trying to convince P's wife and son that P actually was an evil man. Unbelievable, they had fallen for the malicious lies of Q and come to hate P. Sadly, from that time on, P's wife had an affair with Q. Q also passed vile, false rumours to all P's friends to damage P's reputation. All P's friends believed Q's lies. P was completely unaware of this. When he came back, he still lived with his family and still treated Q as a good friend. And also he was still very proud of his 'good' reputation. However, ... although P does not actually **experience** this misfortune, we would judge that he was severely harmed by this event. It is very clear that he lost at least his

reputation and the loyalty of his wife. And both the reputation and the loyalty of his wife are in his interests.

Can a person be harmed in the absence of experiencing harm? In each of the two situations above (plus the five additional cases below), a person is harmed without experiencing harm. Accordingly, we report the following finding: A person can be harmed without experiencing harm.

Can a person be harmed after death? Above, I showed that a person can be harmed without experiencing harm: There is no "Experience Requirement." Below I will show that a person can be harmed after death: There is no "Existence Requirement." First I will now proceed with one more case explicitly showing that a person (call him "Mr. Holiday") can be harmed in the absence of experiencing harm:

(CASE: No Bad News -- by Jeff McMahan)[18]
- [Mr. Holiday is] on holiday on a remote island. Back home, on Friday, his life's work collapses. But, because of the inaccessibility of the island, the bad news does not arrive until the following Monday. On the intervening Sunday, however, the man [Mr. Holiday] is killed by a shark; so he never learns that his life's work has come to nothing.

This is yet another example showing that no "Experience Requirement" is necessary in order for a person to be harmed. But, nevertheless, is an "Existence Requirement" necessary in order for a person to be harmed? The following cases show that a person can be harmed after death:

(CASE: When Killed -- by Jeff McMahan)[19]
- On reflection, it seems hard to believe that it makes a difference to the misfortune he [Mr. Holiday] suffers whether the collapse of his life's work occurs shortly before he is killed or shortly afterward. Yet, according to the Existence Requirement, this difference in timing makes **all** the difference. If the collapse of his life's work occurs just before he dies, then, even though he never learns of it, he suffers a terrible misfortune. If, on the other hand, it occurs just after he dies, he suffers no misfortune at all. If we find this hard to believe then we may be forced to reject the Existence Requirement.

(CASE: No Burial -- by George Pitcher)[20]
- Bill Brown promises his dying father that he will bury him in the family plot when he dies. Bill instead sells his father's corpse to a medical school for dissection by students.

(CASE: Disgruntled Neighbor -- by George Pitcher)[21]
- In World I, a philosopher spends his entire life working on a metaphysical system that he believes to be, and desperately wants to be, the Truth about reality. And it is! After his death, his system is universally accepted, endlessly discussed, and he is acclaimed as the greatest philosopher who ever lived. World II is exactly the same as World I up to the time of the philosopher's death; but in this world, a disgruntled neighbor burns the philosopher's house down the day after his death, and his writings are destroyed. We may imagine that he never revealed his metaphysical views to anyone; so his system is irretrievably lost, and the philosopher is remembered only by a few friends and the hostile neighbor. We would all, I think, judge that the philosopher's life in World I is better than his life in World II, and that the neighbor's vicious action in World II really harms the philosopher.

(CASE: Forges Documents -- by Joel Feinberg)[22]
- Suppose that after my death, an enemy cleverly forges documents to "prove" very convincingly that I was a philanderer, an adulterer, and a plagiarist, and communicates this "information" to the general public that includes my widow, children, and former colleagues and friends. Can there be any doubt that I have been harmed by such libels?

The four cases we have just presented above show that a person can be harmed after death. Just as there is no "Experience Requirement" necessary in order for a person to be harmed, likewise there is no "Existence Requirement" necessary in order for a person to be harmed. Accordingly, we report the following finding: A person can be harmed after death.

Previously we found that a person can be harmed without experiencing harm. We also found that a person can be harmed after death. Moreover, it is obvious that death can impair the objective interests of a person. If harm is the impairment of objective interests, then -- contrary to Epicurus -- death **can** be a harm to the person who dies.

The term **death** in the present context is often defined something like this: The permanent (irreversible) cessation (end) of life, existence, or consciousness.[23] I now point out, however, that even given our present philosophic intent and context, death defined as permanent or irreversible is not without its problems. For one thing, as I have indicated elsewhere, what is deemed permanent or irreversible may be seen as relative to the state of empirical learning (the level of our science-technology).[24] Moreover, if empirical tests necessarily involve empirical corroboration or empirical refutation (either or both), then **"permanent death"** is in principle potentially open to eventual refutation but not to eventual corroboration.[25] In other words, death viewed as a temporary condition that is potentially reversible by far-future science-technology (**"temporary death"**) is open to empirical corroboration in the far-future but is not open to empirical refutation.

Is it possible that the set of all **"permanently"** dead persons can be (or, using far-future science-technology, can be made to become) a **null** set? Is it possible that the set of all **"temporarily"** dead persons can include (or, using far-future science-technology, can be made to come to include) **all** dead persons? It seems that both logically and empirically the answer to both questions is yes. Moreover, this answer apparently applies not only to people and the set of all persons, but also to worlds and the set of all universes.

If our future world, even the very structure of the cosmos itself, is subject to modification by persons -- then we may want to take some empirical or physical theories more seriously than others based on the fact that dead persons have objective interests. We have reported above our finding that it is both logically and empirically possible that the set of all **"temporarily"** dead persons can include (or, using far-future science-technology, can be made to come to include) **all** dead persons. A next step is to more specifically consider alternative empirical theories and the reasonableness of a paradigm shift. Isaac Newton's empirical theory or paradigm would seem to be somewhat friendly to our project of making death temporary. Moreover, beyond this, Albert Einstein's relativistic paradigm would seem to be somewhat friendly to our project of making death temporary. One may say, based on an Einstein-like approach, that the future (since it is not past) is not fully determined and thus even in principle cannot be reliably scientifically predicted in detail. But the past (since it is past) has been fully determined and thus in principle can be reliably scientifically retrodicted in detail. Karl Popper explains Einstein this way: "Thus ... according to special relativity, the past is that region which can, in principle, be known; and the future is that region which, although influenced by the present, is always 'open': it is not only unknown, but in principle not fully knowable ... The predictions demanded by 'scientific' determinism must be interpreted, from the point of view of special relativity itself, as **retrodictions**."[26]

The three major philosophic definitions of **harm to persons** are explored by Jack Li.[27]   Li's conclusions about harm identify **personhood** not with desires or with goods but with objective interests. Both his analysis and my different analysis below find that harm to persons (alive or dead) involves the impairment of their objective interests:

1) Is harm to persons **the thwarting or frustration of desires**?   But dead people are "experiential blanks"[28] and have no desires. Dead people have no sensations, experiences, hopes, or fears. Yet, contrary to Epicurus, we have found above that dead people can be harmed. Moreover, sometimes a particular desire can harm our advancement toward ultimate personhood. Thus, objectively, the thwarting of such a desire would be good or beneficial rather than bad or harmful.

2) Is harm to persons **the deprivation of goods**?   But dead people have no goods in that they have no life and no liberty. Dead people cannot pursue happiness or act to achieve goals or dreams. Yet, contrary to Epicurus, we have found above that dead people can be harmed. Moreover, in this context, "goods" seems more ambiguous and less accurate than "objective interests." Sometimes a particular "subjective interest" (e.g., a particular "desire" or a particular "good") can harm our advancement toward ultimate personhood. Thus, objectively, the thwarting, deprivation, or impairment of such a "subjective interest" or "desire" or "good" would be beneficial rather than harmful.

3) Is harm to persons **the impairment of objective interests**? Persons can indeed be harmed without experiencing harm; moreover, persons can indeed be harmed after death. A dead person is a (dead) person; every person (alive or dead) has objective interests. Considerations above thus suggest the following, a new principle of personhood: "Once a person, always a person."

Jack Li, following Joel Feinberg and John Kleinig, differentiates **subjective** interest ("X is interested in Y") from **objective** interest ("Y is in X's interests").[29]   But unfortunately Li then goes on to follow Feinberg and Kleinig further:[30] (A) "Y is in X's [objective] interests" equals "X has a **justifiably claimed** stake in Y"; and, (B) "X has a stake in Y" equals "X is likely to gain or lose from Y ... " Below I show that assertions (A) and (B) are seriously flawed.

(A) Here the phrase "justifiably claimed" is presumably used in order to differentiate objective interests from merely subjective interests (such as certain desires or goods that are not in our objective interests). A problem with "justifiably claimed," however, is that our objective interests remain our objective interests whether or not we "claim" them. Likewise, our objective interests remain our objective interests whether or not we "justify" them.

(B) Here the phrase "likely to gain or lose" is used. It is perhaps natural to think of our objective interests as somehow connected to gaining or losing. But in fact our objective interests remain our objective interests whether or not some gain or lose is "likely" or unlikely, more probable or less probable.

It may be assertions A and B that unnecessarily cause Li to consider three challenges to his definition of harm as serious enough to warrant some convoluted thinking to defend himself. The three challenges are as follows:[31]

- A crime boss invests large amounts of money and energy in an attempted bank robbery, but his attempt is foiled. According to your definition [Li's definition of harm], it seems that the crime boss is harmed.

- I buy a lottery ticket, but I do not win one million dollars. Had I won one million dollars, I would have benefited, but my failure to benefit shouldn't be regarded as harmful. However, it seems that according to your definition [Li's definition of harm], winning one million dollars is in my interests, so I am harmed.

- I have an annual salary of one hundred thousand dollars ... and my employer ... fails to give me a raise, ... [so it seems that according to Li's definition I am harmed].

I find it obvious that the "solution" to these alleged "challenges" is simply to explain the ordinary meaning of "impairment of objective interests" (including the rejection of faulty assertions A and B, which we have already done). In order to make things even more clear, if such is needed, I point out that here impairment is to **persons** (persons have objective interests); thus it would be clearer to think of persons advancing toward their objective interests, including advancement in their ethical learning. Presumably Robin Hoods are good or ethical, and Crime Bosses are bad or unethical -- even if they each attempt, successfully or unsuccessfully, to rob banks. Thus we may suppose that it is in the objective interest of Robin Hoods to successfully rob banks and that it is in the objective interest of Crime Bosses to be unsuccessful or be caught if they attempt to rob banks.

We now have a clearer definition of harm: Harm to persons is the impairment of their **advancement** toward their objective interests, including the advancement of their **ethical** learning. So indeed we do not say a person is harmed because utopia is not achieved in the next three seconds. Perhaps over a course of four or five seconds things will be no worse than presently.

If not before, it should now be obvious that the three alleged "challenges" require no "solution." It is in the objective interest of the Crime Boss to fail in the bank robbery (or possibly to be caught by police). It may or may not be in my objective interest to win the lottery or gain a higher salary (including the additional headaches and responsibilities this may entail). But failure to secure utopia in the next three seconds is not necessarily an impairment (major setback) to the advancement of my objective interests.

Previously we found, contrary to Epicurus, that death **can** be a harm to the person who dies. We also found that "harm is the impairment of objective interests" -- but went on to clarify this further. To wit: Harm to persons is the impairment of **advancement** toward their objective interests, including advancement of their **ethical** learning.

It may be faulty assertions A and B (see above) that contribute to Li's claim "that the death of an elderly person who has led a full and worthwhile life is not a great misfortune for him."**[32]** On the contrary, our analysis above tells us that being "elderly" (in the sense of age-related disability) and being "dead" (mortal) are generally not in the objective interests of persons. In general, becoming disabled or being mortal does not contribute to an optimal "never-ending" journey of a person toward ultimate personhood. In the following example by Feinberg, Li makes modifications in brackets to support his "one-century" view:

- Thus, if I have an annual salary [life] of one hundred thousand dollars [100 years], and my employer [God] gives me a fifty thousand dollar [50 year] raise, I benefit substantially from this largesse. If he [God] fails to give me a raise, I am not so benefited, but surely not harmed either…If he [God] reduces me to five thousand [50 years]…however, he [God] not merely fails to benefit me, he [God] causes me harm…**[33]**

Li's analogy does not hold up. First of all, God is love (not our harmful or helpful employer) and wishes us to take the initiative and to self-advance toward ultimate personhood. Such an adventure in discovering and advancing one's objective (ethical and other) interests will take much longer than a mere one-century. Secondly, life is not like a mere job or salary. If one is alive and healthy, one may be able to obtain another job or salary. But death ends one's life and life-plan; one does not then obtain another life or life-plan. Beyond this, self-improvement and world-betterment are in our objective interests. Nature (not God) indifferently causes events like drought, earthquake, crop failure, smallpox, AIDS, cancer, age-related disability, and death. Via the advancement of our objective empirical interests, we learn to regulate nature; via the advancement of our objective ethical interests, we turn the world from indifference into love.

Endnotes

[1] Jack Li, **Can Death Be a Harm to the Person Who Dies?** (Dordrecht, The Netherlands: Kluwer Academic Publishers, 2002), 67-97 & 157-167.

[2] Ibid., 68.

[3] Kenneth E. Boulding, **Ecodynamics: A New Theory of Societal Evolution** (Beverly Hills, CA: Sage Publications, 1978 & 1981), 139-140.

[4] Ibid., 338-339.

[5] Bertrand Russell and F. C. Copleston, "The Existence of God," in **Classical and Contemporary Readings in the Philosophy of Religion**, 2nd ed., ed. John Hick (Englewood Cliffs, NJ: Prentice-Hall, 1970), 293.

[6] Ibid., 294.

[7] Many of my remarks related to this question are based on, or inspired from, my reading of: Soren Kierkegaard, **Works of Love**, trans. Howard Hong and Edna Hong (New York: Harper & Row, Torchbooks, 1964).

[8] Ibid., 23

[9] Ibid.

[10] See N. F. Fedorov, chapter nine of the present volume.

[11] Li, 79.

[12] Jeff McMahan, "Death and the Value of Life," in **The Metaphysics of Death**, ed. John Martin Fischer (Stanford, CA: Stanford University Press, 1993), 255.

[13] Thomas O. Buford, **Personal Philosophy: The Art of Living** (New York: Holt, Rinehart and Winston, 1984), 164; also see 165. Buford himself does not mention how long a life-span has been or should be.

[14] Epicurus, "Letter to Menoeceus," in **Philosophic Classics: Volume I: Ancient Philosophy**, ed. Walter Kaufmann (Englewood Cliffs, NJ: Prentice Hall, 1994), 392-395.

[15] Li, 69.

[16] Ibid., 28.

[17] Ibid., 28-29.

[18] McMahan, 235.

[19] Ibid., 240-241.

[20] George Pitcher, "The Misfortunes of the Dead," in **The Metaphysics of Death**, ed. John Martin Fischer (Stanford, CA: Stanford University Press, 1993), 160.

[21] Ibid., 163.

[22] Joel Feinberg, "Harm to Others," in **The Metaphysics of Death**, ed. John Martin Fischer (Stanford, CA: Stanford University Press, 1993), 180-181.

[23] John Martin Fischer, "Introduction: Death, Metaphysics, and Morality," in **The Metaphysics of Death**, ed. John Martin Fischer (Stanford, CA: Stanford University Press, 1993), 3-8.

[24] Charles Tandy, **Doctor Tandy's First Guide To Life Extension And Transhumanity** (Palo Alto, CA: Ria University Press, 2001), 279-286.

[25] John Hick, **Philosophy of Religion** (Englewood Cliffs, NJ: Prentice-Hall, 1963), 101.

[26]   Karl R. Popper, **The Open Universe: An Argument for Indeterminism** (London: Routledge, 1956 & 1991), 61.

[27]   Li, 6-7.

[28]   Fischer, 4.

[29]   Li, 68.

[30]   Ibid.

[31]   The three cases are quoted in Li, 71-72.

[32]   Li, 81.

[33]   Ibid.

Bibliography

Blackmore, S., 1993. **Dying to Live: Near-Death Experiences**. Buffalo, NY: Prometheus Books.

Boulding, K. E., 1978 & 1981. **Ecodynamics: A New Theory of Societal Evolution**. Beverly Hills, CA: Sage Publications.

Braddock, G., 2000. "Epicureanism, Death, and the Good Life," **Philosophical Inquiry** 22, no. 1-2.

Buford, T. O., 1984. **Personal Philosophy: The Art of Living**. New York: Holt, Rinehart and Winston.

Camus, A., 1991. **The Rebel: An Essay on Man in Revolt**, Bower, A., trans. New York: Vintage Books.

Epicurus, 1994. "Letter to Menoeceus," in Kaufmann, W., ed., **Philosophic Classics: Volume I: Ancient Philosophy**. Englewood Cliffs, NJ: Prentice Hall.

Feinberg, J., 1993. "Harm to Others," in Fischer, J. M., ed., **The Metaphysics of Death**, Stanford, CA: Stanford University Press.

Feldman, F., 1991. "Some Puzzles About the Evil of Death," **The Philosophical Review** 100, no. 205-27; reprinted in Fischer, J. M., ed., **The Metaphysics of Death**, Stanford, CA: Stanford University Press, 1993.

Feldman, F., 1992. **Confrontations with the Reaper**. New York: Oxford University Press.

Fischer, J. M., 1993. "Introduction: Death, Metaphysics, and Morality," in Fischer, J. M., ed., **The Metaphysics of Death**, Stanford, CA: Stanford University Press.

Fischer, J. M., ed., 1993. **The Metaphysics of Death**. Stanford, CA: Stanford University Press.

Glover, J., 1977. **Causing Death and Saving Lives**. Harmondsworth: Penguin Books.

Green, M. and Winkler, D., 1980. "Brain Death and Personal Identity," **Philosophy and Public Affairs** 9, 105-133.

Hick, J. H., 1963. **Philosophy of Religion**. Englewood Cliffs, NJ: Prentice-Hall.

Kierkegaard, S., 1964. **Works of Love**, Hong, H., & Hong, E., trans. New York: Harper & Row, Torchbooks.

Li, J., 2002. **Can Death Be a Harm to the Person Who Dies?** Dordrecht, The Netherlands: Kluwer Academic Publishers. {J. Li = Jack Li = Jack Lee}

Luper(-Foy), S., 1987. "Annihilation," **The Philosophical Quarterly** 37, no. 148, 233-52; reprinted in Fischer, J. M., ed., **The Metaphysics of Death**, Stanford, CA: Stanford University Press, 1993.

McMahan, J., 1988. "Death and the Value of Life," **Ethics** 99, no. 1, 32-61; reprinted in Fischer, J. M., ed., **The Metaphysics of Death**, Stanford, CA: Stanford University Press, 1993.

Nagel, T., 1979. "Death," in Nagel, T., **Mortal Questions**. Cambridge: Cambridge University Press.

Nagel, T., 1986. **The View From Nowhere**. Oxford: Oxford University Press.

Nozick, R., 1981. "On the Randian Argument," reprinted in Paul, J., ed., **Reading Nozick**. Totowa, NJ: Rowman & Littlefield.

Parfit, D., 1984. **Reasons and Persons**. Oxford: Clarendon Press.

Perry, J., ed., 1975. **Personal Identity**. Berkeley: University of California Press.

Pitcher, G., 1984. "The Misfortunes of the Dead," **American Philosophical Quarterly** 21, no. 2, 217-225; reprinted in Fischer, J. M., ed., **The Metaphysics of Death**, Stanford, CA: Stanford University Press, 1993.

Pitcher, G., 1989. "Epicurus and Annihilation," **Philosophical Quarterly** 39, no. 154, 81-90.

Popper, K. R., 1956 & 1991. **The Open Universe: An Argument for Indeterminism**. London: Routledge.

Rosenberg, J., 1983. **Thinking Clearly About Death**. Englewood Cliffs, NJ: Prentice-Hall.

Russell, B., and Copleston, F. C., 1970. "The Existence of God," in Hick, J., ed., **Classical and Contemporary Readings in the Philosophy of Religion**, 2nd ed., Englewood Cliffs, NJ: Prentice-Hall.

Silverstein, H., 1980. "The Evil of Death," **Journal of Philosophy** 77, no. 7, 401-424; reprinted in Fischer, J. M., ed., **The Metaphysics of Death**, Stanford, CA: Stanford University Press, 1993.

Strawson, P. F., 1959 & 1964. **Individuals**, London: Routledge.

Tandy, C., 2001. **Doctor Tandy's First Guide To Life Extension And Transhumanity**, Palo Alto, CA: Ria University Press.

Unamuno, M., 1913 & 1972. **The Tragic Sense of Life in Men and Nations**, Kerrigan, A., trans. Princeton: Princeton University Press.

Williams, B., 1973. "The Makropulos Case: Reflections on the Tedium of Immortality," in Williams, B., **Problems of the Self**. Cambridge: Cambridge University Press.

Yourgrau, P., 1987. "The Dead," **Journal of Philosophy** 86, no. 2, 84-101; reprinted in Fischer, J. M., ed., **The Metaphysics of Death**, Stanford, CA: Stanford University Press, 1993.

Acknowledgements

I thank those who commented on earlier drafts of "Toward A New Theory Of Personhood." I especially thank Jack Li {Jack Lee} for his particularly insightful critical comments.

Chapter 3
N. F. Fedorov and the Common Task

"N. F. Fedorov And The Common Task: A 21st Century Reexamination" was first published in 2003 and is here reprinted by permission.

Welcome to Volume 1 of the Death And Anti-Death Series By Ria University Press. {*Death And Anti-Death, Volume 1*} Please contact me if you would like to be an editor or a contributor to a future volume.[1] The present volume is in honor of N. F. Fedorov.

In the present article, I will first of all make a few remarks about N. F. Fedorov and his philosophy of "the common task". After that I will in effect reexamine his general project by looking toward future science and technology from our present perspective, the early 21st century. The nine basic sections to the article are entitled: 1. Fedorovian Philosophy; 2. Computers And Transhumans; 3. Feinberg's Law And World Catastrophe; 4. Expanding Beyond Our Cradle Earth; 5. Molecular Nanotechnology; 6. Suspended Animation And Transmortality; 7. Scientific Universal Resurrection; 8. Time Travel; and, 9. Prediction And Retrodiction.

1. Fedorovian Philosophy

Nikolai Fedorovich Fedorov (other transliterations are known, such as Nicholas Fyodorovich Fyodorov) was a Russian librarian, teacher, and philosopher. Bastard born, recent Fedorovian scholarship has established his date of birth as June 9, 1829; his date of death was December 28, 1903. Today he is perhaps best known for two things, his philosophy of "the common task" and that he was the mentor of the great Konstantin Tsiolkowsky.

Fedorov "advocated the ethical priority of a research and development project he called 'the common task,' by which he meant the universal physical resurrection of the dead by future advances in science and technology."[2] Universal active-creative unconditional love was the motivation for his project. The 19th century Danish philosopher Soren Kierkegaard, in his *Works of Love*, reminded us that one "can be deceived in believing what is untrue, but on the other hand, one is also deceived in not believing what is true."[3] "To cheat oneself out of love is the most terrible deception; it is an eternal loss for which there is no reparation."[4]

It may not be altogether unreasonable to compare "the common task" point of view of Fedorov with "the original position" point of view of John Rawls. Rawls ends his great 20th century classic in political philosophy, *A Theory of Justice*, with these words:

> Thus what we are doing is to combine into one conception the totality of conditions that we are ready upon due reflection to recognize as reasonable in our conduct with regard to one another ... Without conflating all persons into one but recognizing them as distinct and separate, it enables us to be impartial, even between persons who are not contemporaries but who belong to many generations. Thus to see our place in society from the perspective of this position is to see it *sub specie aeternitatis*: it is to regard the human situation not only from all social but also from all temporal points of view. The perspective of eternity is not a perspective from a certain place beyond the world, nor the point of view of a transcendent being; rather it is a certain form of thought and feeling that rational persons can adopt within the world. ... Purity of heart, if one could attain it, would be to see clearly and to act with grace and self-command from this point of view.[5]

As we consider Fedorov's "common task" a century after his death, let us examine the future course of science and technology from the perspective of the early 21st century. To be sure, during the last hundred years there have been numerous scientific-technological developments unforeseen by Fedorov -- yet I doubt that in general he would be surprised. But perhaps he would be surprised by one thing: Today some philosophers and scientists believe that artificial intelligence (AI) is possible even to the extent of computers becoming self-conscious (the question, if it is, of "actual mentality" versus "simulated mentality").

2. Computers And Transhumans

Is Syntax Sufficient For Semantics? Whether philosopher John Searle's claim that "syntax is not sufficient for semantics" is correct has important implications for the tasks of systems theorists and model builders.[6] Syntax may be defined as grammar or the way expressions are put together to form proper sentences; semantics may be defined as the meaning of words or utterances. In the context I will discuss below, the distinction between syntax or grammar and semantics or meaning becomes a distinction between the simulation of mentality and actual mentality. Searle engages in a thought experiment to make his point, known as the Chinese Room.

The Chinese Room. I do not know the Chinese language, but I am put in a room or box with detailed instructions for how to supply answers in Chinese to questions put to me in Chinese. I do not understand the questions or the answers, but I carefully follow the instructions. My grammar is correct even though I do not understand the meaning of the words or discourse. Based on this thought experiment, my conclusion is that the room or box does indeed have a program that works -- yet the box has no mentality, no understanding.

Mentality Versus Simulation. To me, it appears obvious that Searle's claim is correct.[7] In other words, I conclude that software, no matter how sophisticated, is not enough to give today's computers the quality or characteristic we call mentality or mind. Such a computer would lack mentality but yet might be able to simulate mentality. Perhaps actual mentality is rooted, at least in part, in something like feeling or motivation or meaning or understanding.

Mentality Involves Feelings? For philosophic arguments to the effect that mentality involves feelings, I refer you to the work of Suzanne K. Langer and of E. A. Burtt.[8] Let me suggest that physical things like stones or chairs and vegetable things like squash or carrots lack mentality or feeling. Or at least most of us would see the feelings of a carrot or squash as of a lower order below animals.

Mentality Involves The Flow of Electrons? The metabolism of plants is of a chemical nature. But the metabolism of animals also involves electricity or the flow of electrons. This suggests to me that perhaps we should concentrate more on the **electronic** aspects (as distinguished from the algorithmic software aspects) of the computer (develop a joint electronic-software ecology or system?) if we want it to have or be a mind rather than simulate mentality.[9]

Wetware Versus Subatomic-Particle (Electronic?) Computers. John Searle believes that wetware (biology) is necessary to mentality. This belief may be correct, but I consider it to be only a guess. My own guess may also be incorrect, but would permit mentality in some computers and not others: The computer would have to be an **electronic** computer, but not just any electronic computer would do (see above). Beyond this, I suppose we could experiment using subatomic-particles other than electrons.[10] (Does mentality necessarily involve the proper flow of electrons or of sub-atomic particles, as distinguished from mere computer software, no matter how sophisticated?)

Construction Of New Mentalities, Values, Emotions. Let us assume for the moment that my highly speculative hypothesis is correct. In such case, might we not be able to construct many different kinds of mentalities with many different kinds of emotions or motivations? Presumably many of these new thoughts and new feelings would be beyond our present ability to comprehend or even imagine.

The Black-White Room. We spoke of the Chinese Room. Now we speak of the Black-White Room. This room too is known to philosophers engaged in thought experiments. Imagine an extremely intelligent person raised completely in an environment, a room, made up (in terms of colors) **only** of the colors black and white. This person studies and learns everything there is to know about the colors of the rainbow.

Do You See What I Hear? Does this person have anything more to learn about color by stepping outside the Black-White Room? Further, we may imagine the person in the Black-White Room as being deaf from birth but knowing everything there is to know about sound and voice. Such thought experiments suggest a need for humility on the part of systems theorists and model builders.

The Deaf Intellectual: A Call For Humility. The thought experiments I have associated with the Black-White Room give rise to an important epistemic insight or question. Until you actually hear what I hear, then what you see, no matter how sophisticated and knowledgeable, will not be what I hear. A sophisticated and knowledgeable deaf person may be able to simulate hearing -- but the experience of actual hearing is nevertheless missing and thus the knowledge is incomplete.

Transhuman Beings With New Mentalities, Values, Emotions. By creating new mentalities, new emotions, new beings I call **transhuman**, we would be able to advance our learning in directions never previously imaginable. Calling humans **trans-apes** is not very enlightening. Likewise the term **trans-human** leaves much unsaid, as we know not what to say.

3. Feinberg's Law And World Catastrophe

Feinberg's Law. Yet regardless of whether our speculations above prove correct or incorrect, it seems that a transhuman transcivilization is likely -- barring catastrophe. According to Feinberg's (so-called) "Law," almost any technology we can imagine which does not contradict the known laws of science will eventually become possible.[11] Moreover, according to Gerald Feinberg, many things which contradict our scientific laws as presently constituted will also come to pass.

Barring Catastrophe. We have spoken of the future of technology with the qualifying phrase **barring catastrophe**. It has been said that the dinosaurs became extinct due to an ecological disaster because they did not have a space program. If they had populated extraterrestrial space, then the Earth-bound disaster would not have extinguished their species. Is there any doubt but that in the long run most of our offspring will be living somewhere in the universe other than on planet Earth? Earth was our cradle, but we cannot live in the cradle forever.

4. Expanding Beyond Our Cradle Earth

ETC Technology For Survival And Flourishing. We have already suggested that one or more technologies will eventually become available to create transhuman beings with transhuman values and emotions. But the space technology I will now talk about is **already** within our grasp and thus available rather immediately if we choose to use it. I refer you to the work of the late Gerard K. O'Neill of Princeton University.[12] ETC technology or the creation of self-sufficient Extra-Terrestrial Communities, large Earth-like greenhouses in space, is already feasible. Moreover, the **self-sufficient** ETCs have the ability to **reproduce at an exponential rate**.

5. Molecular Nanotechnology

Universal Wealth And Leisure? Next I mention molecular nanotechnology, defined by K. Eric Drexler as building things from the ground up atom by atom.[13] This is how a fertilized egg develops into an adult gerbil or human or whale. Nano-scale robots, computers, and factories (too small to be seen by the unaided human eye) will presumably mean an end to human labor and the reign of universal wealth and leisure.

6. Suspended Animation And Transmortality

Then there is suspended animation. As pointed out by Robert Ettinger and Paul Segall, this will allow one-way time travel (so-called) into the future.[14] In the common sense view, time flows in only one direction (forward); a difference here is that present biomedical technology may be unable to prevent the imminent death of our loved ones, but suspended animation may allow access to future biomedical technology, thus perhaps restoring our dying loved ones to full health. Indeed, Ettinger and Segall envision an era of transmortality, a day when all disease, including the disease of aging, has been conquered.

7. Scientific Universal Resurrection

Resurrection Of All Dead Persons. More difficult to envision is the scientific universal resurrection of all dead persons who have ever lived.[15] But, as emphasized by Fedorov, this task is nevertheless a moral imperative which should have begun yesterday. The task appears to be both extremely difficult and extremely ethically compelling.

8. Time Travel

Genuine Two-Way Time Travel. Finally, in closing, I mention time travel.[16] Philosophers now tell us that there appears to be no **logical** inconsistency here so long as we realize that certain laws will constrain our behavior. Anything and everything in our time travel adventures will not be possible. Consider the following analogy: The space travel machine we call an automobile will not transport us under the ocean or to the planet Mars.

9. Prediction And Retrodiction

Retrodiction Versus Prediction. Perhaps -- if I may be allowed to speculate wildly -- perhaps time travel into the past will permit us to observe but not interfere. Yet even this would be a tremendously important learning experience. Too, it might be useful to Fedorov's "common task," our project of scientifically resurrecting all dead persons who have ever lived. Indeed, perhaps retrodiction of the past will prove easier and more important than prediction of the future.[17]

Bibliography

Barrow, J. and Tipler, F. *The Anthropic Cosmological Principle*. Oxford: Oxford University Press, 1988. Copyright 1986. Corrected edition, 1988.

Berdyaev, N. "N. F. Fyodorov." *The Russian Review* 9 (1950) 124-130. Fedorov's thought was not without influence on Berdyaev's existentialism.

Berdyaev, N. *The Russian Idea*. New York: Macmillan Co., 1948. Fedorov and other original Russian thinkers are discussed.

Boulding, K. *Ecodynamics*. Beverly Hills: Sage Publications, 1981. Copyright 1978. Revised edition, 1981.

Boulding, K. *The Meaning of the Twentieth Century*. New York: Harper & Row, 1965. Copyright 1964. First Harper Colophon edition, 1965.

Bova, B. *Immortality: How Science Is Extending Your Life Span - And Changing The World*. New York: Avon Books, 1998.

Broderick, D. *The Spike*. Victoria, Australia: Reed Books Australia, 1997.

Burkhardt, A. (ed.). *Speech Acts, Meaning and Intentions*. Berlin: Walter de Gruyter Co., 1990.

Burtt, E. *In Search of Philosophic Understanding*. New York: New American Library, 1965.

Cetron, M. and Davies, O. *Cheating Death: The Promise and the Future Impact of Trying to Live Forever*. New York: St. Martin's Press, 1998.

Crandall, B. and Lewis, J. (eds.). *Nanotechnology: Research and Perspectives*. Cambridge, MA: MIT Press, 1992.

Deutsch, D. *The Fabric of Reality: The Science of Parallel Universes - and Its Implications*. New York: Penguin, 1997.

Dewdney, C. *Last Flesh: Life in the Transhuman Era*. Toronto: HarperCollins, 1998.

Drexler, K. *Engines of Creation*. New York: Anchor Press/Doubleday, 1987.

Drexler, K. *Nanosystems*. New York: Wiley, 1992.

Ettinger, R. *The Prospect of Immortality*. Garden City, NY: Doubleday, 1964.

Faloon, W. *Disease Prevention and Treatment*. Fort Lauderdale, FL: Life Extension Foundation, 2001.

Fedorov, N. *Filosofiya Obshchago Dela: Stat'i, Mysli, i Pis'ma Nikolaia Fedorovicha Fedorova*, ed. V. A. Kozhevnikov and N. P. Peterson, 2 vols. Originally published by Fedorov's friends and followers after his death, 1906, 1913; reprint London: Gregg Press, 1970. [In Russian.]

Fedorov, N. "The Question of Brotherhood or Kinship, of the Reasons for the Unbrotherly, Unkindred, or Unpeaceful State of the World, and of the Means for the Restoration of Kinship" in Edie, J. M.; Scanlan, J. P.; Zeldin, M.; and Kline, G. L., eds. *Russian Philosophy*. Chicago: Quadrangle Books, 1965. 16-54. This is one place to begin if you want to read Fedorov directly (in English translation).

Fedorov, N. *Sobranie Sochineniy*, 4 vols. + supp. Moscow: Traditsiya, 2000. [In Russian.]

Fedorov, N. *What Was Man Created For? The Philosophy of the Common Task*. Koutiassov, E. and Minto, M. (eds.). Lausanne, Switzerland: Honeyglen/L'Age d'Homme, 1990. From Fedorov's posthumous publications which were in Russian, 1906 and 1913 [see above].

Feinberg, G. *Solid Clues*. New York: Simon & Schuster, 1985.

Finch, C. *Longevity, Senescence, and the Genome*. Chicago: University of Chicago Press, 1990.

Flew, A. *A Dictionary of Philosophy*. London: Pan Books, 1979.

Freitas, R. *Nanomedicine Volume 1: Basic Capabilities*. Austin, TX: Landes Bioscience, 1999.

Gruman, G. *A History of Ideas About the Prolongation of Life*. Salem, NH: Ayer, 1977. Reprint of 1966 edition. [2003 reprint now available from Springer.]

Kosko, B. *The Fuzzy Future: From Society and Science to Heaven in a Chip*. New York: Harmony Books/Random House, 1999.

Kurzweil, R. *The Age of Spiritual Machines*. New York: Viking, 1999.

Langer, S. *Philosophy in a New Key*. New York: New American Library, 1951.

Lepore, E. and Van Gulick, R. (eds.). *John Searle and His Critics*. Oxford: Basil Blackwell, 1991.

Lewin, R. *Complexity: Science on the Edge of Chaos*. New York: Macmillan, 1992.

Lossky, N. *History of Russian Philosophy*. New York: International Universities Press, 1951. Fedorov is included in this history.

Lukashevich, S. *N. F. Fedorov (1828-1903): A Study in Russian Eupsychian and Utopian Thought*. Newark: University of Delaware Press, 1977. The methodology used in this study may not insure full appreciation of Fedorov's thought, but it does demonstrate that his thought was indeed a detailed, coherent philosophy in which the various pieces fit together.

Moravec, H. *Mind Children: The Future of Robot and Human Intelligence*. Cambridge, MA: Harvard University Press, 1988.

Nicolis, G. and Prigogine, I. *Exploring Complexity*. New York: W. H. Freeman and Company, 1989.

O'Neill, G. *The High Frontier: Human Colonies in Space*. New York: Morrow, 1977. Copyright 1976. A recent (year 2000) reprint contains updated information and a CD-ROM.

Pence, G. *Who's Afraid of Human Cloning?* Lanham, MD: Rowman & Littlefield, 1998.

Perry, R. *Forever for All: Moral Philosophy, Cryonics, and the Scientific Prospects for Immortality*. Parkland, FL: Universal Publishers, 2000.

Rawls, J. *A Theory of Justice*. Revised edition. Cambridge, Massachusetts: The Belknap Press of Harvard University Press, 1999. [Original edition was 1971.]

Regis, E. *Great Mambo Chicken and the Transhuman Condition*. Reading, MA: Addison-Wesley, 1990.

Schmemann, A. (ed.). *Ultimate Questions: An Anthology of Modern Russian Religious Thought*. New York: Holt, Rinehart and Winston, 1965; reprint Crestwood, NY: St. Vladimir's Seminary Press, 1977. Selections (translations) from Russian religious thinkers, including Fedorov, concerned with eschatology or other "ultimate" questions.

Searle, J. *Minds, Brains and Science*. Cambridge, MA: Harvard University Press, 1984. Based in part on: Searle, J. "Minds, Brains and Programs." *The Behavioural and Brain Sciences* 3 (1980): 417-57.

Segall, P. *Living Longer, Growing Younger*. New York: Times Books, 1989.

Silver, L. *Remaking Eden: How Genetic Engineering and Cloning Will Transform the American Family*. New York: Avon, 1998.

Silverstein, A. *Conquest of Death*. New York: Macmillan, 1979.

Soloviov, M. "The 'Russian Trace' in the History of Cryonics," *Cryonics* 16:4 (4th Quarter, 1995) 20-23. Closing paragraph describes author's then-current (post-cold-war) and perhaps unprecedented efforts promoting cryonics and immortalism in the former Soviet Union; the article itself acknowledges a debt to Fedorov.

Tandy, C. (ed.). *Doctor Tandy's First Guide To Life Extension And Transhumanity*. Palo Alto, CA: Ria University Press, 2001.

Tandy, C. (ed.). *The Philosophy of Robert Ettinger*. Palo Alto, CA: Ria University Press, 2002.

Tandy, C., and Perry, R., [Fedorov article] in: The **Internet Encyclopedia of Philosophy** at the following place: <http://www.utm.edu/research/iep/f/fedorov.htm>.

Teilhard de Chardin, P. *The Phenomenon of Man*. Wall, B. (trans.). New York: Harper & Row, 1959.

Tipler, F. *The Physics of Immortality: Modern Cosmology, God, and the Resurrection of the Dead*. New York: Doubleday, 1994.

Toffler, A. *The Third Wave*. New York: Bantam Books, 1980.

Waldrop, M. *Complexity: The Emerging Science at the Edge of Order and Chaos*. New York: Simon & Schuster, 1992.

Walford, R. *Maximum Life Span*. New York: Norton, 1983.

Young, G. *Nikolai F. Fedorov: An Introduction*. Belmont, MA: Nordland Publishing Company, 1979. Not only an excellent introduction, but a mine of references and information inviting further Fedorovian research, including Russian language works, many of which are not yet translated (or not fully translated) into English.

Zakydalsky, T. *N. F. Fyodorov's Philosophy of Physical Resurrection*. Ann Harbor, MI: UMI, 1976. A Ph.D. dissertation (Bryn Mawr) of 531 pages. Bibliography has a list of Russian language works.

Zenkovsky, V. *A History of Russian Philosophy*. New York: Columbia University Press, 1953. Fedorov is included in this history.

Endnotes

[1] You may email me at <tandy@ria.edu>.

[2] C. Tandy and R. Perry, "The Anti-Death Philosophy Of N. F. Fedorov" in: C. Tandy (ed.), *The Philosophy of Robert Ettinger*. Palo Alto, CA: Ria University Press, 2002. Page 189.

[3] S. Kierkegaard, *Works of Love*. H. Hong and E. Hong (trans.). New York: Harper & Row, Torchbooks, 1964. Page 23.

[4] Ibid.

[5] J. Rawls, *A Theory of Justice*. Revised edition. Cambridge, Massachusetts: The Belknap Press of Harvard University Press, 1999. Page 514.

[6] J. Searle, *Minds, Brains and Science*. Cambridge, MA: Harvard University Press, 1984. Based in part on: J. Searle, "Minds, Brains and Programs." *The Behavioural and Brain Sciences* 3 (1980): 417-57.

[7] What is **mind** or **mentality**? What is **freedom of the will** or **free agency**? These are major issues in the long history of philosophy. Today they remain unresolved. Most folks would say that "thinking like a computer" is only part of the kinds of things we associate with human mentality, that computers are better than humans when it comes to mechanical or computational processing.

Computer programs are fundamentally algorithmic. But this may not be so for human minds. Although human minds engage in computational or algorithmic processing, this is not what most human minds perceive to be their essence. Humans tend to associate human mentality with a variety of additional processes/states as well, not the least of which is seen as involving emotion, sensitivity, perception, meaning, understanding, inwardness/subjectivity, and the exercise of free agency. Algorithmic processing may succeed or fail in its task. But a human mind "inwardly feels" and may freely choose to do good or evil; humans typically continue to believe this moral/amoral distinction even when they find that their choices were in fact far less free than they had previously supposed.

Thus John Searle may be seen as defending the common sense or default position. For me, his Chinese Room thought experiment is rather straightforward. Few people really believe that the Chinese Room or their abacus or desktop computer has mentality. Although most folks (including me) are intuitively convinced that the Chinese Room is mindless, they agree that it is good at simulating (giving the outward appearance of) at least part of human mentality called natural language. Yet despite our inability to articulate a comprehensive, integrative definition of mentality that we find satisfying, most of us would agree that an insect unable to engage in natural language has more mentality than the mindless Chinese Room.

In this paper I speculate or offer ideas that may or may not prove correct or fruitful. According to Searle, given our present dismal state of knowledge, we have little choice but to speculate; such speculation is a function of philosophic reflection. See, for example, E. Lepore and R. Van Gulick (eds.). **John Searle and His Critics**. Oxford: Basil Blackwell, 1991. Page 341.

[8] Langer's eyesight and health deteriorated before she could complete to her satisfaction in its entirety **Mind: An Essay on Human Feeling**. It is based on her previous work, especially S. Langer, **Philosophy in a New Key**. New York: New American Library, 1951. For another approach, see E. Burtt, **In Search of Philosophic Understanding**. New York: New American Library, 1965.

[9] So far as I am aware, I am the first person to explicitly articulate this notion.

[10] So far as I am aware, I am the first person to explicitly articulate this notion.

[11] The esteemed physicist wrote too for the general public. See, for example G. Feinberg, *Solid Clues*. New York: Simon & Schuster, 1985.

[12] G. O'Neill, *The High Frontier: Human Colonies in Space*. New York: Morrow, 1977. Copyright 1976. A recent (year 2000) reprint contains updated information and a CD-ROM.

[13] K. Drexler, *Engines of Creation*. New York: Anchor Press/Doubleday, 1987; K. Drexler, *Nanosystems*. New York: Wiley, 1992.

[14] R. Ettinger, *The Prospect of Immortality*. Garden City, NY: Doubleday, 1964; P. Segall, *Living Longer, Growing Younger*. New York: Times Books, 1989.

[15] N. Fedorov, *What Was Man Created For? The Philosophy of the Common Task*. E. Koutiassov and M. Minto (eds.). Lausanne, Switzerland: Honeyglen/L'Age d'Homme, 1990. From Fedorov's posthumous publications which were in Russian, 1906 and 1913; G. Young, *Nikolai F. Fedorov: An Introduction*. Belmont, MA: Nordland Publishing Company, 1979; T. Zakydalsky, *N. F. Fyodorov's Philosophy of Physical Resurrection*. Ann Harbor, MI: UMI, 1976; F. Tipler, *The Physics of Immortality: Modern Cosmology, God, and the Resurrection of the Dead*. New York: Doubleday, 1994.

[16] D. Deutsch, *The Fabric of Reality: The Science of Parallel Universes - and Its Implications*. New York: Penguin, 1997; R. Perry, *Forever for All: Moral Philosophy, Cryonics, and the Scientific Prospects for Immortality*. Parkland, FL: Universal Publishers, 2000.

[17] As I have said, this matter is wild speculation. Yet given certain assumptions, assumptions not viewed as wild by many current philosophers, it (the quest for scientific resurrection) is a moral imperative. Retrodiction, so it would seem, deserves considerably more serious philosophic and scientific study than it has received so far.

Chapter 4
Unburying the Dead

> "Unburying The Dead: Posthumous Harms And Posthumous Benefits -- A Solution To The Missing Subject Problem" was first published in 2003 and is here reprinted by permission.

Previously we found, contrary to Epicurus, that death **can** be a harm to the person who dies. We also found that "harm is the impairment of objective interests" -- but went on to clarify this further. To wit: Harm to persons is the impairment of **advancement** toward their objective interests, including advancement of their **ethical** learning.[1]

Persons can indeed be harmed without experiencing harm; moreover, persons can indeed be harmed after death. A dead person is a (dead) person. Considerations above thus suggest the following, a new principle of personhood: "Once a person, always a person."[2]

When is one harmed -- and who is the subject of the harm -- if one's harm is posthumous (i.e., if the harm is to a "dead person")? Do any of the proposed solutions seem totally coherent and convincing? In the absence of being able to make clear sense of my new principle of personhood ("Once a person, always a person"), I was tempted to give in to Li's view.[3] But fortunately for my new principle, Catterson came along.[4]

Typically when we think of a **person**, we think of a person with a healthy body and a bright mind. Yet sometimes we meet persons with unhealthy bodies and unbright minds. But have we ever met a **person** without a body or without a mind? Even a ghost has a (ghost kind of) body. And a statue has no mind. If ghosts exist, we would say they are persons (even if their bodies and minds function somewhat differently from yours and mine). Certainly medical mannequins, "talking" dolls, and cremated ashes exist -- yet we do not call them **persons**.

Is the term "dead person" an oxymoron? My answer, following Catterson, is **no**. Dead persons exist now as real facts about the past. The past is a fixed (fully determined) unity that **will always exist** even if the universe dies. This is one reason why scientists and philosophers say that time travel into the past is in principle possible but that changing the past is impossible even in principle. (However time branching seems logically possible.)

In chapter 16 (this volume) {*Death And Anti-Death, Volume 1*} Catterson presents a plausible account of time where the past exists in the present. Indeed, Catterson argues (page 423) that "there can be no coherent conception of an A-series that posits the passing away of the past ... thus there could be no possible world where the dead do not exist." Accordingly, both dead persons and living persons have no choice but to be presently existing. A presently existing dead person may be characterized as person-identity (fully complete and up-to-date) **fact-information** that: 1) lacks a mind that functions as a living person at the present moment (e.g., is an "experiential blank"); and, 2) lacks a body that functions as a living person at the present moment. These two "lacks" correspond to two aspects of posthumous harms: a person can be harmed without experiencing harm and a person can be harmed even if not presently alive.

It is false that a dead person does not presently exist **at all**. Thus there is no "missing subject" problem. It is the (presently existing) dead person (fact-information, not cremated ashes) who is the subject of (present) posthumous harms and benefits.

Moreover, it is possible that the death of a person is not permanent. As John Hick points out, "permanent death" **cannot** be empirically corroborated in a finite period of time; but "permanent death" **can** be empirically refuted in a finite period of time (and in individual cases has in some sense been refuted often via CPR or advanced medical techniques).[5] In addition, this logical principle applies not only to the death of persons but also to the death of universes.[6] (To empirically show that universe U will never be resurrected, we would have to wait forever.)

If the popular "big bang" theory (including that space-time had a beginning) is correct, then presumably most of what is real lies in the future (not in the present or the past). Thus our present and past experiences (such as: "Nothing is certain but death and taxes") probably do not constitute a very good representative sample of reality. However the fact that person P really existed in the past as a living person means that it really is empirically possible for person P to exist as a living person. (This is something that cannot be said of fictional- or fantasy-characters.) Based on considerations above (including my Introduction to the present volume) {*Death And Anti-Death, Volume 1*} -- scientists, philosophers, and other living persons, in advancing their objective interests (including their empirical, artistic, and ethical learning) should love all dead persons and take into account the objective (real) interests of dead persons.

Unburying the dead -- resurrecting all dead persons by scientific means -- may not yet be at hand. Indeed, developing a beloved community consisting of all the living and all the dead will offer great challenges -- but also great benefits. As we look toward the future, as we consider the vast varied ranges and regions of reality ahead, it would seem arrogant in the extreme not to engage with reality in Fedorov's sacred "common task," the beloved community.[7]

Bibliography

Hick, J. H., 1963. **Philosophy Of Religion**. Englewood Cliffs, NJ: Prentice-Hall.

Tandy, C., ed., 2002. **The Philosophy Of Robert Ettinger**. Palo Alto, CA: Ria University Press.

Endnotes

[1] Tandy, 179.

[2] Ibid., 177.

[3] Chapter 15 in the present volume by Jack Li. {Chapter 15 of *Death And Anti-Death, Volume 1*}

[4] Chapter 16 in the present volume by Troy T. Catterson. {Chapter 16 of *Death And Anti-Death, Volume 1*}

[5] Hick, 101.

[6] Tandy, 175.

[7] I thank Steven Luper for his comments on a previous draft of the present paper, thus saving me from at least one grievous error.

Chapter 5
Is the Universe Immortal?

"Is The Universe Immortal?: Is Cosmic Evolution Never-Ending?" was first published in 2004 and is here reprinted by permission.

Welcome to Volume 2 of the Death And Anti-Death Series By Ria University Press. {*Death And Anti-Death, Volume 2*} Please contact me if you would like to be an editor or a contributor to a future volume. [1] The present volume is in honor of Immanuel Kant and Alan Turing.

{This paragraph about the chapter contributions to *Death And Anti-Death, Volume 2* has been omitted.}

In the present article, I will introduce you to some of the thoughts of Eric J. Chaisson as expressed in his 2001 book, *Cosmic Evolution: The Rise of Complexity in Nature*. Chaisson is a physicist and systems scientist of considerable status. His book attempts to integrate several scientific disciplines in an apparently conservative, rather than wildly speculative, way. Yet the general integration he produces results in potentially important new ideas/knowledge (at least ideas and formulations new to me).

Chaisson explains why there is good reason to believe that the universe may be immortal. Chaisson explains why there is good reason to believe that cosmic evolution may be never-ending. In short, it seems that our universe is increasing in *both* (!) disorder/entropy/randomness *and* order/negentropy/information. Chaisson attempts to explain in physical- and systems-science terms how this putatively paradoxical situation works and how the evolution of complexity can be never-ending.

This five-part paper describes Eric J. Chaisson's interdisciplinary account of the rise of complexity in nature, *Cosmic Evolution*. Part one asks, "Is the universe a deterministic machine?" explaining Chaisson's negative answer. Part two, "Is cosmic evolution a good personal philosophy to live by?" takes issue with Chaisson's positive response. Part three, entitled "three thought experiments -- and their surprising results when scientifically tested" recounts a bit of history not covered by Chaisson. Part four explains Chaisson's formulation of "free energy rate density" as a first step toward characterizing the engine or fuel of cosmic evolution. Part five explains why Chaisson believes a never-ending perpetually-dynamic universe should not be ruled out.

1. Is The Universe A Deterministic Machine?

Chaisson reports that the universe is not mechanistic (the "idea that all natural processes are machines, explainable in terms of Newtonian mechanics and thus ultimately predictable"). [p. 234] [2] And he defines determinism as follows: "The idea that all events have specific, definite causes and obey precise, natural laws, making their outcomes completely predictable; from any particular initial state, one and only one sequence of future states is possible." [p. 231] Neither mechanism nor determinism seems applicable to the universe. Nevertheless, the present is rooted in the past.

> Everything is what it is because it got that way.
> ■ Kenneth E. Boulding, *Ecodynamics*

> Hydrogen is a light, odorless gas which, given enough time, changes into people.
> ■ Anonymous humorist
> [Quoted in Eric J. Chaisson, *Cosmic Evolution*]

> At any moment any atom may deviate slightly from its course.
> ■ Epicurus
> [See Rex Warner, *The Greek Philosophers*]

According to a number of interdisciplinary scientists, including Chaisson: "Nature is not clean and clear ... chance mixes with necessity, reductionism with holism, physics with biology ... [to give us] an uncertain future." [p. 223] "In quantum physics, there is no single final state but only several possible [probabilistic] alternative states." [p. 35] Moreover, "chance behavior of individual particles in the microdomain can surprisingly yield observable changes in the macrodomain." [p. 36] Indeed, "a portion of the water in a kettle atop a fire could (theoretically) freeze while the remainder boils, although this is so unlikely as to be considered impossible in our practical experience." [p. 45] But over great spans of time, strange (unlikely) things happen, thus effecting the future course of the universe.

2. Is Cosmic Evolution A Good Personal Philosophy To Live By?

Although presenting a more refined analysis, Chaisson emphasizes two great changes following the beginning of spacetime: The transformation of the Radiation Era into the Matter Era, and of the Matter Era into the Life Era. With the Matter Era, the universe contained not only energy in the form of radiation but also energy in the form of matter (atoms). In the Life Era, we find a third form of energy -- conscious intelligence as advanced cultural-technological civilization.

The Life Era thus does not begin with biology or photosynthesis, but with advanced civilization "because only with the origin of technologically manipulative life, not just life itself, does life exert leverage over both radiation *and* matter." [p. 145] Undirected evolution seems to be giving way to conscious or directed evolution of the universe. Thus, according to Chaisson, we now need a good personal philosophy of life, a new worldview based on science and technology to guide the future evolution of the universe well rather than poorly. Advanced civilizations have the capacity to change their hereditary nature (e.g., genetic engineering) as well as their physical, biological, and cultural environments. Whether there are advanced civilizations beyond Earth (conscious intelligences already with some limited capacity to protect, destroy, improve, change, or direct the course of the universe) is presently unknown.

But although Chaisson may be serving up a delicious dish of integrative science, it seems clear to me he has not (yet) articulated a robust, or even bare outline for a, philosophy of life. He seems to believe he will be able to follow through with his "new philosophy" now under construction. As he looks at the big-bang beginning through his quantum lens, asking "Why is there something rather than nothing?" he is convinced that "science now seems poised to address the origin of the primal energy of creation itself." [p. 6]

Whether or not this turns out to be the case, Chaisson has a huge amount of work to do in other directions. For example, about all he has to say so far (in this book) about metaphysics is contained in a single sentence: "Decidedly, modern science, with its physical laws and symmetry rules expressed in mathematical language, is not inconsistent with the Platonic ideal of a deeper reality of 'eternal ideas and unchanging forms.'" [p. 220] Similarly, about all he has to say so far (in this book) about moral philosophy is as follows: "Is humankind part of a cosmological imperative, heading, perhaps with other sentient beings, toward some astronomical destiny? Put bluntly yet magnanimously, the scenario of cosmic evolution grants us unparalleled 'big thinking,' from which may well emerge the global ethics and planetary citizenship likely needed if our species is to remain part of that same cosmic-evolutionary scenario." [p. 224]

3. Three Thought Experiments -- And Their Surprising Results When Scientifically Tested

For reasons articulated above, I conclude that Chaisson is misguided if he believes he is offering a new philosophy of life (which is his claim). On the other hand, science generally (including the integrative science of cosmic evolution) has provided us with vital new knowledge and will no doubt continue to do so. To show just how vital (and surprising) such new knowledge can indeed be, let me take you back in time to review three scientific developments.

By simple thought experiment, Aristotle concluded that heavy bodies fall faster than light bodies. A more refined thought experiment by Galileo led to a contrary conclusion. When the experiment was finally tested empirically, the results showed what Galileo had expected. Heavy bodies do not fall faster. Thus we credit Galileo with establishing the law of uniform acceleration for falling bodies. If I had been betting, I would have bet against Galileo; I would no doubt demand the experiment be run again, as the results just *had* to be wrong! (But then I always thought the empirical evidence favored the hypothesis that the Earth does not move: I feel no movement when I stand still and no great wind sweeps me off my feet!)

Now let us move from the early 17th century to the late 19th century. (I will reformulate the details, but hopefully not the substance, of the thought experiment to make it more entertaining and straightforward:) We have a new railway system that runs non-stop from Columbia University to Stanford University. The train travels at 200 miles per hour. On the train heading toward California, I throw a baseball at 50 miles per hour toward Stanford University. Question: For an outside observer, how fast is the baseball travelling? What if you said 250 miles per hour but a Dr. Einstein said 50 miles per hour? (Well, the empirical results here would show you right and Einstein wrong.) But what if I could throw something *much* faster than a mere 50 miles per hour -- would the mathematical principle we established still work? (In the case of the baseball example: 200 miles per hour plus 50 miles per hour equals 250 miles per hour.)

Now the train is traveling toward California at one-half the speed of light and instead of throwing a baseball I turn on a flashlight aimed in the direction of Stanford University. For an outside observer, how fast is the light travelling? (Dr. Einstein will supply us an explanation for the surprising Michelson-Morley experimental results.)

Einstein's relativity theory offers an explanation as to why the speed of light is the same for me on the train as it is for the outside observer. (Again, the empirical results about the speed of the *baseball* make sense to me, but without an understanding of relativity theory, the empirical results about the speed of *light* seem to me ridiculous. Run the experiment again! After all, I recall that in high school chemistry class, my experimental results often came out contrary to the textbook!)

I would have bet against the law of uniform acceleration for falling bodies. I would have bet against relativity theory. (Please don't mention that I would have bet against the hypothesis that the Earth moves!) Next we will find Einstein, in the 20th century, "betting" that, as he said, "God does not play dice." Dr. Albert Einstein was referring to quantum theory. In his mind, surely a deterministic theory of quantum mechanics would have to replace the statistical interpretation. But the empirical results proved embarrassing for Einstein and corroborating for the probabilistic theory known today as quantum physics.

In 1935, Einstein, along with (Boris) Podolsky and (Nathan) Rosen, articulated a thought experiment (known as the "EPR" experiment); it sought to establish that quantum physics (a non-deterministic or statistical theory) was surely flawed ("incomplete" is the technical term). The EPR thought experiment showed that if quantum physics were correct, some surprising/unacceptable/bazaar results would follow. Subsequent empirical tests corroborated the "predicted" results, to Einstein's embarrassed astonishment.

I have discussed the thought experiments turned empirical experiments for a reason. The integration of science to produce a scientific account of cosmic evolution can be done in a "speculative" way for future (perhaps far future) empirical results to corroborate or belie. But Chaisson, instead, uses the established or standard theories to get coherent new knowledge about the story of cosmic evolution. He emphasizes that no new science (theory) is needed to understand cosmic evolution. (I am tempted to compliantly remark that standard quantum physics is already strange enough without having to make it even more bazaar.) So here goes some interesting details of cosmic evolution as integratively elucidated by Chaisson ...

4. Expansion Of The Universe And Free Energy Rate Density

The early (before matter) universe was in equilibrium, meaning that actual entropy equaled maximum possible entropy. But the expansion of the universe created a new environment, thus allowing the possibility of matter. Once the equilibrium symmetry was ruptured, the actual entropy in the expanding universe "increases less rapidly than the maximum possible entropy ... Accordingly, the expansion of the Universe can be judged as the ultimate source of order, form and structure, promoting the evolution of everything in the cosmos." [p. 130]

So far as I am aware, Chaisson is the first person to come up with a promising way to characterize order throughout cosmic evolution and to compare complexity in one system with another. He explains his term, "free energy rate density" but ultimately concludes he has much more work to do to refine the term further. Previous thinkers have suggested terms like: information; meaningful information; know-how; knowledge; knowledge capacity; learning advancement capacity; negentropy (negative entropy). He explains why his term, while imperfect, is more functional than the others. E.g., even if we knew how to measure either negentropy or entropy, it "holds true only for static, equilibrium conditions" thus leaving out non-equilibrium systems. (Note: Perhaps there is no such thing as an absolutely equilibrium system within the "real" world; systems within the universe always have an environment to potentially interact with?)

The "free energy rate density, alternatively called the specific free energy rate" is "expressed in units of energy per time per mass." [p. 134] A "galaxy clearly has more energy than a cell, but of course galaxies also have greater sizes and masses; it is the organized energy *density* that best characterizes the degree of order or complexity in any system." [p. 134] "Free energy rate density" is the "amount of energy (available to do work) flowing through a system per unit time and per unit mass." [p. 233]

Chaisson's term "is familiar to astronomers as the luminosity-to-mass ratio, to physicists as the power density, to geologists as the specific radiant flux, to biologists as the specific metabolic rate, and to engineers as the power-to-mass ratio." [p. 134] The interdisciplinary term refers to "a rate of energy flow, not a flux per se, and a mass density, not a volume density." [p. 134]

According to a "free energy rate density" analysis, the human body is more complex than the Earth's biosphere but less complex than the human cranium. The human brain is less complex than modern civilization, suggesting at least to me the possible limited capacities of hermits and isolated communities. But Chaisson points out that without further refinement of his term, the term too often fails to accurately compare the complexity of one system to another. Nevertheless, something like "free energy rate density," according to Chaisson, should be seen as the engine or fuel of cosmic evolution.

5. Will The Universe End?

> Order, we must have order -- but not too much order!
> ■ A. N. Whitehead
> [Quoted in Eric J. Chaisson, *Cosmic Evolution*]

"Darwinian biological evolution and Lamarckian cultural evolution, among other catalytic effects during physical evolution, might well have fostered additional complexity on local scales beyond that possible by universal expansion alone. That's probably why, in a relative sense, physical evolution is sluggish, biological evolution moderate, and cultural evolution rapid." [pp. 141-2]

Chaisson's analysis implies that entities superior to human and human-civilizational intelligence already exist or will exist somewhere in the universe. And a modest scientific search for extraterrestrial life -- rather than mere armchair philosophizing -- seems justified. Chaisson also suggests we should advance beyond his "free energy rate density" oversimplification -- e.g., "to examine more closely how, and how well, open systems utilize their free energy flows to enhance complexity. ... Very low energy flows mean the system will" tend to equilibrate "with the thermal sink, whereas very high flows will ... damage the system to the point of destruction. ... It's a little like the difference between watering a plant and drowning it." [p. 144] Supernovae "are clearly objects of extremely high energy rate density precisely as expected for one of the most unconstructive events in Nature." [p. 144]

There is no guarantee that conscious intelligence or advanced technological civilization will prove victorious over matter as matter was victorious over radiation in the early universe. E.g., our civilization may fail because it never gains "control of material resources on truly galactic scales. ... [On the other hand, it may well be that] the longevity of technological civilizations everywhere is inherently small." [p. 145]

The future is unpredictable in that it does not follow Ockham's Razor: Improbable things happen. Apparently "change is the hallmark for the origin, maintenance, and fate of all things." [p. 2] The expanding "Universe self-generates a thermal gradient, and increasingly so with time, suggestive of an ever-powerful heat engine ... To be sure, we must emphasize throughout the statistical nature of all these processes, meaning that the growth of order is not a foregone conclusion, nor is the universe a machine." [pp. 128-9]

The classical (deterministic) physicists of old used to predict an ultimate state of eternal cold rest for the universe. But "a more modern analysis is not so dire, suggesting that the maximum possible entropy will likely never be attained. In an expanding Universe, the actual and maximum entropies both increase, yet not at the same rate; a gap opens between them and grows larger over the course of time, causing the Universe to increasingly depart from [an end-of-the-universe scenario.] ... We need not be so pessimistic, indeed it is this inability of the cosmos to ever reach true maximum disorder that allows order, or lack of disorder, to emerge in localized, open systems." [p. 29] Indeed, Chaisson concludes, a "perpetual [!] stream toward richness, diversity, and complexity, the outcome of which cannot be foreseen, may be the true fate of the Universe." [p. 219]

Bibliography

Boulding, Kenneth E., 1981 [first edition, 1978]. *Ecodynamics: A New Theory of Societal Evolution*. Beverly Hills, California: Sage Publications.

Chaisson, Eric J., 2001. *Cosmic Evolution: The Rise of Complexity in Nature*. Cambridge, Massachusetts: Harvard University Press.

Warner, Rex, 1958. *The Greek Philosophers*. New York: New American Library.

Endnotes

[1] You may email me at <tandy@ria.edu>.

[2] Page numbers in brackets refer to Eric J. Chaisson's *Cosmic Evolution: The Rise of Complexity in Nature*.

Chapter 6
Earthlings Get Off Your Ass Now!

"Earthlings Get Off Your Ass Now!: Becoming Person, Learning Community" was first published in 2004 and is here reprinted by permission.

> We are being smothered by people who believe themselves to be absolutely right, whether because of their machines or their ideas. And for all who cannot live except in dialogue and friendship with men, this silence is the end of the world.
> ■ Albert Camus, *Actuelles I*

This paper is about the present age and how to prevent doomsday. Part One is entitled "an age of absurdity and uncertainty." The present age as absurd and uncertain is suggested by historical events specified herein.

Part Two is entitled "an age of anti-absurdity and uncertainty." Certainty is absurd; uncertainty is anti-absurd. It is not wise to put all of humanity's eggs (futures) into one basket (biosphere). Construction of large, comfortable, permanent, self-sufficient Extra-terrestrial Green-habitat Communities (EGCs, not to be confused with space stations) is feasible this century. Ratification of a proposed Space Treaty now, before we weaponize space, would prevent an arms race in outer space, our future home.

--PART ONE--
AN AGE OF ABSURDITY AND UNCERTAINTY

According to a number of cultural/intellectual historians of the 20th century Western world, it is not unreasonable to characterize our present age (perhaps about 90 years old and counting) as a time of absurdity and uncertainty. [1] Some 19th century artists and intellectuals may be seen as prophets of the absurdities and uncertainties of the following century and beyond. But in the Western world generally, the late 19th century was filled with self-confidence: Westerners thought that humans represented the leading edge of life on earth, and that Western civilization was destined to be humanity's great leader.

The present age (the 20th century and beyond) may be characterized as a time of discontinuities, pluralities/diversities, absurdities, and uncertainties -- as suggested by the following sets of historical events:

- World Wars One And Two
- Events Between The Wars And After The Wars
- 20th Century Developments In Science
- 20th Century Developments In Mathematics
- 20th Century Developments In Psychology
- The Early 20th Century Modernist Movement
- 20th Century Existential Literature And Absurd Drama
- The 20th And 21st Century Postmodernist Movements

World Wars One And Two

Many of the late 19th century believed that human progress was inevitable. This included, for example, belief that, over time, wars would decrease in intensity, scope, and duration. Indeed, perhaps Western civilization had already devised a continental system guaranteeing nothing more severe than limited battles or small wars between European nations.

World War One (1914-1918)

The war that began in 1914 only gradually turned into a great "World War." The war started in an almost careless mood with full-dress parades. But World War One proved to be larger than any previous war in European history.

Approximately 74 million were mobilized by all sides in the "Great War." The war was also unprecedented in another respect: This war involved entire societies, not merely soldiers. The concept of "total war" was born.

World War One resulted in the collapse of traditional empires, and, from the wreckage, new nations arose. Moreover, to Europe's further embarrassment, it appeared Europe had been unable to settle its own affairs (end the long war) without intervention from the New World (the USA). Europe and the West no longer seemed worthy as the leading edge of world progress, leadership, and perfection.

World War Two (1939-1945)

"The war to end all wars" failed too in that a Second World War began in 1939. The war of 1939, far more than the war of 1914, was a **world** war. (Thus, it seems progress is **not** inevitable.)

Events Between The Wars And After The Wars

The "**Roaring Twenties**" (1920s) were called "roaring" because of the exuberant, freewheeling popular culture of the decade following World War One. The Roaring Twenties were a time when many people indulged in new or illegal styles of dancing, dressing, and behavior, and rejected many traditional moral standards. This decade is also known in the United States as the "**Jazz Age**," marked by increased popularity of ("wild") jazz (music), and by attacks on convention in many areas of American life. Many Americans defied Prohibition (the outlawing of alcoholic beverages nationwide from 1920 to 1933). The nickname "flappers" was given to young women in the 1920s who defied convention by their dress and by such behavior as drinking and smoking in public.

Throughout the 1930s there was a "**Great Depression**," a worldwide economic crisis that continued into World War Two. Indeed, World War Two and its aftermath left the world in an unprecedented crisis situation. The attempted genocide of the Jews, the large scale death camps (gas chambers), and the use of the atomic bomb in World War Two raised terrifying questions about the extreme possibilities of human inhumanity.

There were Germans who listened to their Bach or read their Goethe after a day's work at the gas chambers (death camps). The Americans not only initiated development of the Manhattan Project, but actually authorized use of the atomic bomb -- using atomic mass death and destruction not against military targets, but against cities of civilians. Indeed, the possibility of instant extermination through nuclear war or weapons of mass death and destruction makes many traditional values and conventions seem obsolete.

20th Century Developments In Science

In addition to nuclear weapons, additional science related developments radically altered our views of the universe and of ourselves. The atomic bomb was a kind of "test" (so to speak) of some of the ideas of Albert Einstein. With Einstein's new physics the dividing line between mass and energy was now far from clear. His theory of relativity says that space, time, and motion are not absolute (as Newton had assumed), but relative to the observer.

Previously, the intelligent layperson could read and understand the great thinkers like Newton and Darwin. But the thoughts of Einstein and other scientists were now only available to specialized experts. Moreover, the new physics seemed to postulate a world without continuity or absolutes, a world in which nothing was certain. A world in which scientific laws are given names like "relativity" (Einstein) or "indeterminacy"/"uncertainty" (Werner Heisenberg) is a world in which nothing is as it appears to be to the human senses.[2]

20th Century Developments In Mathematics

If physical reality itself is inherently "relative" and "indeterminate" or "uncertain," then surely we can at least depend on mathematics ("the language of the sciences") to give us nonabsurdity and certainty? But the work of Kurt Gödel and others in the 20th century showed that even mathematics is less secure and dependable than had been assumed. Gödel, often cited as the greatest logician of the 20th century, is credited with proving two theorems that must be bad news for anyone wanting to construct a theory that will tell the whole truth.

Gödel's First Incompleteness Theorem states that for any consistent logical system able to express arithmetic, there must be true sentences within the system that are undecidable (cannot be proved true) within the system. Gödel's Second Incompleteness Theorem states that no such (consistent) system can prove its own consistency. The moral of the two incompleteness theorems would seem to be that truth inevitably outstrips formal provability.

20th Century Developments In Psychology

Foremost in upsetting traditional notions about human nature was the work of Sigmund Freud. The new psychology proclaimed that it is not unusual for humans to engage in **self-deception**. Freud's theories that our actions and behavior are rooted in our **unconscious**, rather than in our conscious mind, seemed to many a highly **pessimistic** view of human nature. Freudian influences, including the idea of sexual repression, can be found throughout 20th century art and literature. Human nature, reason, and logic are apparently far more dark and fallible than had been thought possible.

The early 20th century West experienced a radical questioning of past traditions. For example, the Western cultural roots of Greek-Roman "Reason", Hebrew-Christian "Religion", and Modern-Progressive "Science" were no longer believable traditions or were of highly uncertain value. It seems neither Reason nor Religion nor Science provided a sure path out of the absurd wilderness.

The Early 20th Century Modernist Movement

Even in the very early 20th century, it seemed to a number of Western artists and intellectuals that the 20th century had broken, in a new and radical way, with previous tradition. A radical break in the Western arts became evident with a series of new styles that can all be loosely grouped under the name "**modernism**." (Many "isms" may be grouped under "modernism." Modernism is the philosophy and practices of "modern art" -- especially a self-conscious break with the past and a search for new forms of expression.)

In the early 20th century West we find:

- Desire for sexual freedom
- Demand for greater freedom for women
- Motion pictures
- Jazz music expressed the fast, frantic, free way of life of the "Roaring Twenties" (1920s)
- Life was speeding up and changing
- The city of Paris (France) was the place to go to exchange ideas and create the new

Almost every serious artist of the 20th century felt the necessity either to react against modernism or to build on its innovations. Most of the great "modernist" writers of the early 20th century spent some time living in Paris (France). A few of the "modernist" literary/artistic "isms" include:

- **Dadaism** is based on deliberate irrationality and negation of traditional artistic values; seeks the fantastic and absurd; life is random and uncontrolled.
- **Surrealism** is related to dadaism; heightened awareness of the conflict between the rational and irrational; produces dreamlike, fantastic, or incongruous imagery or effects. (Two artistic examples: Salvador Dali, painter; Jean Cocteau, filmmaker.)

- **Expressionism** seeks to depict the subjective emotions and responses that objects and events arouse in the artist; German expressionists expressed the deep, hidden drives of human beings -- modern society alienates the individual.
- **Futurism**, initiated in Italy about 1909, sought to give formal expression to the dynamic energy and movement of mechanical processes.

In some ways the popular American writer Ernest Hemingway was an "atypical" modernist. In general, modernist writers and artists experienced a rift between artist and public. The lack of contact with a public, except for a small group of fellow intellectuals, gives the writer or artist greater freedom to experiment but also encourages esotericism.

20th Century Existential Literature And Absurd Drama

The temper of our time is a mixture of anticipation and anxiety, optimism and pessimism. The philosophy of existentialism illustrates this blend of hope and despair. "Existentialism" is said to have been "founded" in the 19th century -- but its popularity dates from the period of the 20th century's World Wars. The Theater of the Absurd represents an extension of existentialist philosophy and literature into drama in the 20th century.

Existentialism

For the existentialist, it is the individual, not the abstract concept of person or humanity, that constitutes true reality. The individual is a **stranger** or an **outsider** in an alien, hostile world. Such loneliness is seen as a call to action and free choice. Two of the 20th century's leading existentialists were Jean-Paul Sartre (1905-1980) and Albert Camus (1913-1960).

In his novel **The Stranger** (or **The Outsider**), and in his essay **The Myth of Sisyphus**, Camus demonstrates his concept of the "absurd" -- the fundamental meaninglessness of human life and traditional beliefs. As Camus said in **The Myth of Sisyphus**: "A world that can be explained by reasoning, however faulty, is a familiar world. But in a universe that is suddenly deprived of illusions and of light, man feels a stranger. His is an irremediable exile, because he is deprived of memories of a lost homeland as much as he lacks the hope of a promised land to come. This divorce between man and his life, the actor and his setting, truly constitutes the feeling of Absurdity." Thus individuals must create their own morality, their own way of resisting or rebelling against the "absurd."

The Theater of the Absurd

Beginning in the 1950s, some dramatists came to experience, as had Albert Camus, a profound sense of absurdity. Samuel Beckett, Harold Pinter, and others shared a pessimistic vision of humanity struggling to find a purpose and to control its fate. Absurdist playwrights did away with the logical structures of traditional theater. The busyness of the characters underscores the fact that nothing happens to change their existence. The ridiculous purposeless behavior and talk of the characters give the plays a sometimes dazzling comic surface, but there is an underlying serious message of metaphysical distress.

The 20th And 21st Century Postmodernist Movements

Sometimes the contemporary **historical period** we have characterized as "a time of discontinuities, pluralities/diversities, absurdities, and uncertainties" is called "post-modernist" or "postmodern." But there are at least two additional ways these terms are sometimes used: In the late 20th century, two drastically different sets of **artistic-literary movements** were each given the same name: "post-modernist" or "postmodern." One of the two sets of movements may be characterized as a reaction against the Modernist artistic-literary movement of the 20th century. Post-modernist or postmodern in this case means revival of traditional artistic-literary elements and techniques.

But the second set of artistic-literary movements is an intensification or yet further evolution of Modernism. In this case the post-modernist or postmodern perspective goes beyond Modernism's aspiration to (a new) unity. The search for unity has been abandoned in favor of pluralities/ diversities of styles and interpretations.

Thus no single cultural tradition or mode of thought can serve as a **metanarrative** (a universal voice, or totalizing story, for all human experience). Jean-Francois Lyotard defines **The Postmodern Condition** as "incredulity toward metanarratives." Incredulity does not mean disbelief -- it means **inability** to believe.

The Modernists of the 20th century felt exiled, like outsiders or strangers, and sought a new unity that they, and perhaps all persons, could believe in. But today many postmodernists **celebrate** incredulity as a liberating, humanizing force. They hear a new key: For them, pluralities of voice are beautiful, not terrifying.

--PART TWO--
AN AGE OF ANTI-ABSURDITY AND UNCERTAINTY

Does this great celebration celebrate uncertainty? In some sense it **does**. Does this great celebration celebrate absurdity? In some sense it does **not**. Does the great celebration embrace anti-absurdity? Apparently it does. A life of celebration and multiple flourishing is not without its values. Its values are those associated with celebration or flourishing and with resisting or rebelling against the "absurdity" of anti-celebration and anti-flourishing. The celebratory kind of incredulity involved here is neither totalizing, absolute, nor nihilistic.

So how do we make the transition from an absurd here and an uncertain now -- to a vital world of celebration and multiple flourishing? Perhaps nourishing a certain kind of uncertainty and of celebration may help transform an absurd uncertain universe into a flourishing uncertain multiverse. Voltaire wrote to Frederick the Great in 1767: "Doubt is not a pleasant condition, but certainty is an absurd one."

Certainty Is Absurd, Uncertainty Is Anti-Absurd

In our history of learning to advance toward personhood and community, dialogue and friendship, we should know by now that one's fundamental beliefs should be held tentatively rather that absolutely. Socrates is said to have been the world's wisest person because he knew that he did not know, whereas everyone else was certain. Yet Socrates was committed to dialogue and friendship even at the risk of his very life.

Martin Luther King, Jr. encouraged our **Strength to Love**, but yet observed: "Nothing in all the world is more dangerous than sincere ignorance and conscientious stupidity." The lesson of dialogue and friendship, of anti-ethnocentrism and anti-certainty, is difficult to learn. But along with such difficult humility comes a sense of responsibility and community. Descartes, engaged in thought, declared: "I am thinking, therefore I am." But Camus, engaged in action, felt: "I am rebelling, therefore we are." In rebelling, in taking responsibility against certainty and absurdity, one feels one is acting not on one's own behalf only, but in the name of all persons.

The Up Project ("UP")

Why get excited about anything that doesn't go straight up?
- A quotation widely attributed to Arthur C. Clarke

Above we have already suggested at least some of the absurdities of the present age. Now we ask what practical measures should be tentatively taken or experimentally attempted to promote celebration and flourishing, dialogue and friendship, anti-certainty and anti-absurdity. "Life has taught us that love does not consist in gazing at each other but in looking outward together in the same direction." (Antoine de Saint-Exupery, *Wind, Sand and Stars*)

In the absence of catastrophe, does it not seem likely that in the long run most of our offspring will be living somewhere in the universe other than on planet Earth? And is it not desirable that persons live in a friendlier or less absurd world than we find ourselves in today? Accordingly, might there be practical actions we can promote today to take us from here to there?

If the dinosaurs had had a space program like the Up Project ("UP"), they would not be extinct. As we ask the question of absurdity and practical measures to be attempted against it, it seems wise to ask questions like the following: Can doomsday be prevented? If so, how? What can we Earthlings do here and now? Can the future be better than the present? If so, how? What can we Earthlings do here and now? Is the surface of planet Earth or is any existing planet really the right place or best location for an expanding technological civilization? If not, what is the alternative -- and what can we Earthlings do here and now?

Multiple Biospheres Are Better Than One

The fact that Earthlings presently exist together in a single biosphere global village is a rather absurd position to be in if we seek to prevent doomsday. If something catastrophic happens to Earth's biosphere, then something catastrophic happens to all Earthlings. It is not wise to put all of humanity's eggs (futures) into one basket (biosphere).

Extra-terrestrial Green-habitat Communities ("EGCs") should not be confused with space stations. We are really talking about two very different entities. Yet twentieth century technology was already sufficiently advanced so that Earthlings could have initiated the Up Project ("UP") if they had chosen to do so. (To be sure, most twentieth century Earthlings were unaware of the opportunity to initiate our first steps toward building large comfortable homes and permanent self-sufficient greenhouse cities in space, EGCs.)

A vital capacity of UP to be realized relatively early-on (in a project of many decades) is that of drastically reducing the cost of launching stuff from Earth into space. According to a world famous physicist now serving as President of the Space Studies Institute, Freeman J. Dyson:[3] "The public is well aware that with present-day launch-costs human activity in space must remain a spectator sport. ... It took fifty years to go from the Wright brothers' Flyer One of 1903 to the modern air-transport system with huge numbers of commercial aircraft flying routinely all over the world." I point out that today's world is a different and speeded-up world -- and that when we explicitly decide to do something (whether build the atomic bomb or land a human on the moon), it tends to meet success comparatively sooner rather than later. Several different approaches to building a public highway system into space have been identified by Dyson as deserving support. Two different systems, one for people and the other for cargo, may provide two separate kinds of public highways into space.

Extra-terrestrial Green-habitat Communities or EGCs can be built from the resources of the moon or the asteroids (either or both). Each EGC would be home for thousands; later EGCs would be even larger (an Extra-terrestrial Green-habitat Community of millions seems feasible). Rotation of the large and spacious greenhouse habitat provides simulated gravity for the people and plants living on the inner surface. Adjustable mirrors provide energy from the sun and simulation of day and night. Sooner or later, the following seems feasible for EGCs:

- Unlimited energy from the sun
- Control of daily weather and sunlight
- Self-sufficient EGCs
- Expansion of self-sufficient EGCs at a geometric rate
- Unlimited free or cheap land via EGCs

Earthlings get off your ass now! The following metaphorical insights have been widely quoted by EGC experts: "The Earth isn't sick, she's pregnant!" "The Earth was our cradle, but we will not live in the cradle forever." "Space habitats [EGCs] are the children of Mother Earth."

Two World Governments Are Better Than None?

Dr. Carol Rosin[4] has argued that achieving an enforceable, permanent ban on space-based weapons is feasible only at this moment in history **before** actual weapons are placed in space. She proposes a carefully worded World Space Preservation Treaty as an effective and verifiable multilateral agreement to prevent an arms race in outer space. This includes prevention of the weaponization of outer space.

The 1967 UN Outer Space Treaty has been signed by 116 nations, banning weapons of mass destruction from outer space. The proposed Space Preservation Treaty establishes and funds the Outer Space Peacekeeping Agency that will monitor and enforce the ban. This Treaty would serve as a catalyst or foundation for a cooperative world space economy, security system, and society.

This innovative approach may shift our collective consciousness toward concern for:

- World health and education
- A clean and sustainable environment
- International security needs through information sharing
- Research and development of clean energy and stimulation of the world economy
- Our role in the infinite universe
- Peace preserved in space as leading to peace on earth

The Treaty can serve to facilitate the building of a world economy fit for the Space Age. This would include a variety of public and private cooperative space ventures not related to space-based weapons. For example, defense activities in space not related to space-based weapons include communications, navigation, surveillance, reconnaissance, early warning, and remote sensing. There is indeed a vital need for such military related activities in space.

With this treaty in place, the solving or management of global problems thus becomes more feasible. By capping the arms race before it escalates into space, we world citizens are transforming the entire weapons mindset and war industry into a cooperative world space industry. As we begin to work in space (and eventually make EGCs our permanent homes for quality living), we will find it in our economic interest to establish in space:

- Factories
- Hospitals
- Hotels and resorts
- Schools and universities

According to Rosin, weapons deployed in space will have the ability to target any point on earth with great accuracy, allowing the nation controlling those weapons to dominate the entire earth with impunity. At present, the war industry thinks it has a mandate to expand into space. Nevertheless the war industry has the ability to change its mind and transform itself in line with the proposed Treaty. For example, satellites have important functions: to monitor the environment, to early-warn us of human-made or natural disasters, and to verify arms agreements.

By living peacefully in space, we will eventually learn to live peacefully on earth. This Treaty will not immediately solve all problems, but it is an unusually important necessary step in the right direction. It offers hope for the future, and opportunities to invest in a future worth living in. Under this Treaty, the military-academic-industrial complex will move into space, but within a framework that enforceably bans space-based weapons and encourages world security and cooperation and the flourishing of multiple biospheres.

Once the proposed Treaty is ratified, an Outer Space Peacekeeping Agency will be established. This agency would not only enforce the proposed Treaty but would enforce the 1967 Outer Space Treaty (for the first time!) as well. The proposed Treaty (including Peacekeeping Agency) will be the international mechanism by which the nations of the world community work together, with effective enforcement, so they can protect themselves against any aggressor nation that might attempt to unilaterally (or with allies) weaponize space.

This monitoring and enforcement applies equally against all nations and parties, whether signatories to the Space Preservation Treaty or not. This Treaty in essence creates a world agency, similar to a United Nations of Space, under a sovereign multilateral treaty establishing a world outer space jurisdictional authority with full enforcement powers. It is not subject to the terrestrial limitations of the Security Council under the United Nations Charter, a prior Treaty that will have been superceded for purposes of jurisdiction in outer space.

Becoming Person, Learning Community

> "Ah, you're a Maximalist," said the beadle. "No, I am only a Minimalist, I merely want the Minimum -- that we save our own lives."
> ■ Israel Zangwill, *Ghetto Comedies*

By now surely we are in the process of learning to cherish the humble or positive qualities of uncertainty, reject the arrogant or negative qualities of certainty, constructively rebel against absurdity, and "save our own lives." As we "Minimalists" implement the Up Project to prevent doomsday, it will have effects some may identify as "Maximalist" -- but with a difference. Indeed, such multiple biospheres (EGCs) will act as catalysts for multicultural flourishing within the context of a world at stable peace.

Given the UP framework, we now have the capacity to move toward a more beautiful personhood and a more advanced community. We can learn how to become a becoming person. We can learn how to become a learning community. Doomsday is not inevitable. Becoming person, learning community -- may yet be within our grasp.

Bibliography

Asimov, Isaac. *Asimov's Chronology of the World.* (HarperCollins Publishers). 1991.

Camus, Albert. *Actuelles I.* (Gallimard). 1950.

Camus, Albert. *The Myth of Sisyphus and Other Essays.* (Vintage Books). 1991.

Camus, Albert. *The Rebel: An Essay on Man in Revolt.* (Vintage Books). 1991.

Camus, Albert. *The Stranger.* (Sagebrush Bound). 1999.

Globus, Al and Yager, Bryan. *Space Settlements.* <www.nas.nasa.gov/Services/Education/SpaceSettlement/>. 2004.

King [Jr.], Martin Luther. *Strength to Love.* (Augsburg Fortress Publishers). 1981.

Lyotard, Jean-Francois. *The Postmodern Condition: A Report on Knowledge.* (University of Minnesota Press). 1984.

O'Neill, Gerard K. [1975 Interview] <http://lifesci3.arc.nasa.gov/SpaceSettlement/CoEvolutionBook/Interview.HTML>.

O'Neill, Gerard K. *The High Frontier: Human Colonies in Space.* (Morrow). 1977. [A year 2000 reprint (from Collectors Guide Publishing, Inc.) contains updated information and a CD-ROM]

Platt, John R. [editor] *New Views of the Nature of Man.* (University of Chicago Press). 1971.

Platt, John R. *The Step to Man.* (John Wiley & Sons Inc.). 1966.

Rosin, Carol. The Institute for Cooperation in Space website at <www.peaceinspace.com>.

Saint-Exupery, Antoine de. *Wind, Sand and Stars.* (Harcourt). 1992.

Zangwell, Israel. *Ghetto Comedies.* (Classic Books). 1907.

Endnotes

[1] E.g., World War One began in 1914. On the other hand, others (such as John R. Platt and Isaac Asimov) think of the present age as beginning with the end of World War Two in 1945 (Asimov) or soon thereafter with the massive advance of scientific research and technological development legitimized by explicit ongoing government support (Platt designates 1950 as year one).

[2] See pages 29-31 of the present volume {*Death And Anti-Death, Volume 2*} for a brief discussion by me of the new physics (i.e., relativity physics and quantum physics). {This is section 3 of Chapter 5 in the present work.}

[3] Personal communication from Freeman J. Dyson (September 9, 2004).

[4] The information in this section is based on Carol Rosin's work presented on the Institute for Cooperation in Space website at <www.peaceinspace.com>.

Chapter 7
Ettinger's 1964 Thesis

> "Ettinger's 1964 Thesis: Indefinitely Extended And Enhanced Life (Immortality) Is Probably Already Here Via Experimental Long-Term Suspended Animation" was first published in 2005 and is here reprinted by permission.

Robert Ettinger begins the Foreword to his 1964 classic thus [p. xxi]:[1] "Most of us now breathing have a good chance [prospect] of physical life after death [immortality] -- a sober, scientific probability of revival and rejuvenation." Moreover, this amazing scientific-technological development "seems to have gone virtually unnoticed." The "prospect of immortality is not idle speculation" but "urgently requires action by all of us as individuals."

Ettinger defines "immortality" as "indefinitely extended life." [p. xxi] He argues that such immortality is practical for us now, not just for our descendants. Ettinger further argues that immortality in its present form -- experimental long-term suspended animation now -- "is desirable from the standpoints both of the individual and of society." [p. xxi]

Below I will use the term "cryonics" even though the term was not invented until 1965-6.[2] I suppose for present purposes it is reasonable enough to associate cryonics with low temperature biostasis or experimental long-term suspended animation now -- as a possible door into a future world far advanced in the technology of life enhancement (a step beyond mere life extension technology). Ettinger refers to this future world as "the Golden Age" and optimistically believes there is a good chance he personally will experience it, first perhaps as a human repaired to youthful health, then as a super-enhanced immortal or superhuman.

Chapter One: Death's Reversibility

Ettinger's "practical guide to immortality" consists of eleven chapters. Chapter one serves as an overview of his analyses and arguments in the chapters to follow. In chapter one, Ettinger finds that we use the term "death" in a variety of ways.

Death defined to give a **religious** meaning will depend on one's religion. Legal death or death defined in a **legal** sense is related to a legal document or a determination of law. Death simply as lack of heartbeat and respiration is **clinical** death; but if current techniques are unable to restore clinical function, then it becomes **biological** death. The "irreversible degeneration" of cells may be labeled as **cellular** death -- yet "the question of reversibility at any stage depends on the state of medical art." [p. 3]

Cryonic hibernation at the temperature of liquid nitrogen puts the biologically dead person in a state that prevents further deterioration and permits possible access to the more advanced medical technology of the future, including the far distant future. With bad luck, the cryonaut will not be revived from biological death. With good luck, revival means living as a superhuman or super-enhanced immortal in a Golden Age. Thus the potential prize is enormous.

Chapter Two: Cooling Down

Activity of the cerebral cortex of a rat ceases at about 18°C. Cooling of rats to a body temperature far below 18°C for over one hour showed persistence of memory in the revived rats. Moreover [p. 25], each memory "seems to be stored in many separate locations in the brain, and therefore may withstand widespread damage;" and memories "consist of [a kind of] chemical coding [that] ... may be hardy and resistant to damage."

Typically many of the cells of a frozen and thawed organ or animal survive the freezing and thawing, even when function cannot be restored to the whole entity by our crude techniques of today. Use of cryoprotective blood substitutes improves matters. If some cells of an organ or animal survive freezing and thawing, perhaps many other cells of the entity almost survive the freezing and thawing or almost survive the freezing stage or do survive the freezing stage (but do not fully survive the thawing). Indeed, there is reason to believe that the thawing (i.e., thawing and post-thawing) stage causes more damage than the freezing stage, especially if cryoprotective agents are used in the controlled freezing.

Thus our task is to preserve persons soon after legal death -- leaving thawing and post-thawing treatments to the future. Future (or far distant future) technology should have substantial ability to infer the state of the original healthy cell from the state of the damaged cell in its frozen context. At room temperature [p.34]: "The period of grace before all of the body cells die is measured at least in hours, and perhaps in days."

Chapter Three: Thawing Out

Many persons clinically dead for a few minutes have been revived to life. Some persons clinically dead for many minutes have been revived to life using even our crude techniques of today. Our biomedical science and techniques related to resuscitation, transplantation, freezing, thawing, and repair will be a little better in the near future and a lot better in the far future.

Chapter Four: Today's Choices

A mere century of living in a merely human body is a beginning, but pales by comparison to the superhuman or super-enhanced immortal you can become. "You can change your mind after freezing, but not after burial." [p. 73] Via legal documents, infrastructure arrangements, and life insurance policies, make provisions now for you and your loved ones to be cryonically hibernated at legal death.

Chapter Five: Religious Issues

Death viewed as reversible and as a matter of degree may seem novel. But revival of the dead is not a new problem. The clinically dead have indeed been revived not only after a few minutes but after many minutes. Anecdotal accounts suggest a few cases where clinical death lasted over two hours before successful revival.

Religious and non-religious folks alike readily accept such scientific accounts of many minutes and anecdotal accounts of two hours of clinical death. Given the recent speed of scientific-technological advance, most folks find no religious or philosophical impossibility with many minutes becoming many years or two hours becoming two millennia. In that sense, the prospect of revival from cryonic hibernation is a scientific (not religious or philosophical) matter. Likewise, the matter of extending healthy lifespan from 50 years to 50 millennia.

Most folks, including most religious folks, sometimes use automobiles for transportation; they would not agree that if "God had intended man to go forty miles an hour, He would have provided him with wheels instead of legs." [pp. 76-77] Indeed, a wide variety of religious and non-religious philosophies agree in asserting that one has a duty or responsibility to seek improvement and betterment, both for oneself and for others. Common beliefs about the wrongfulness of suicide and murder easily extend to end-of-life-cycle cases of failure to freeze oneself (suicide) or failure to freeze others (murder). Such logic would seem to establish cryonic hibernation as the default position or ethical imperative at legal death. For the revived cryonaut, cryonic hibernation was a method of life extension -- and the merely legal death was not real and permanent death.

Cryonic hibernation allows "the present generation to share the longevity which our descendants will have in any case." [p. 87] Such longevity cannot guarantee us certain and literal immortality. But it can mean indefinitely extended and enhanced life for thousands of years -- for you and me already alive today.[3]

Chapter Six: Legal Issues

The **rights** of both animate and hibernating persons must be properly legitimized by law and custom. The **obligations** of both animate and hibernating persons must be properly legitimized by law and custom. "Heretofore a corpse has had in itself neither rights nor obligations; now it will have both." [p. 93] Incompetents exist in relatively small numbers today, but cryonauts "will constitute an enormous body of influence which must be duly recognized and represented." [p. 103]

Eventually, cryonic hibernation or biostasis will become the default legal position at legal death. Generally in such an end-of-life-cycle event, failure to freeze will be illegal, a case of manslaughter. And a sloppy murder will be legally differentiated from a murder allowing ordinary cryonic (biostasis) treatment of the victim.

Debts will be "subject to simple interest only, while assets may accumulate compound interest." [p. 104] Thus you **can** take it with you! Although "it is true that the freezer era will be the era of the Golden Rule, the fraternal outlook will become general only gradually." [p. 104] Hence at some point society will feel obligated to freeze the poor. "For failure to pay the premiums on one's freezer insurance, the death penalty seems a trifle severe." [p. 103]

Chapter Seven: Economic Issues

As we look to the distant future (say, the 23rd century), things look good. Self-improving, self-reproducing computers should eventually lead to the technological equivalent of the magic genie lamp. Our increasing abilities to control and reorganize matter and energy should result in undreamed of real wealth and available energy. And the size of the universe provides a lot of space into which we will expand.

On the other hand, as a practical problem of today, the population explosion is all too real. And with or without cryonics, we will have longer lifespans and then immortality (indefinitely extended lifespans). The long view that comes with cryonics and immortality will help us address population problems with realistic birth control policies.

If overpopulation is still a problem in the 23rd century, then hibernation in the form of perfected suspended animation will be useful. One portion of the population could alternate with others by hibernating for a set period of time. We could honeycomb the earth and other planets to great depth. We could create new planets if desired.

But what are the relatively more expensive per capita costs of biostasis today? What are the economics of experimental long-term suspended animation right now? Consider the cost of cryonic hibernation staff, equipment, and facilities (including liquid nitrogen) over a very long period of social-economic ups and downs. Ettinger finds the cost of hibernation today to be about one order of magnitude above traditional arrangements (funeral, burial, and related expenses). This means that "life" (hibernation) insurance is generally the way to go and is easily affordable by most young adults in North America and Western Europe. Indeed, through the wonder of compound interest, it may be possible to awaken wealthy. To be sure, future governments could decide to limit the wealth of cryonauts.

The advancement of cryonics, suspended animation, longevity, and immortality should, over time, benefit some traditions and not others. The advancement of biostasis would seem to favor "permanence of the family and of institutions, a strengthened feeling of the unity of mankind, [and] an ingrained sense of our endless responsibility for each other." [p. 125] Thus ultimately cryonics and immortality should serve to help humankind avoid temptations like fanaticism and terrorism.

Chapter Eight: Personal Identity

This section, Chapter Eight, is a highly original contribution to professional philosophy. Professional philosophers subsequently duplicated his work without being aware of Ettinger's thought experiments related to biostasis and personal identity.[4] Ettinger presents his philosophic work -- and reports his findings straightforwardly despite their apparent incongruity with the hibernation project.

Previously Ettinger had reported his findings that death is reversible and that death is a matter of degree. Ettinger's thought experiments in this chapter suggest "that individuality is an illusion" [p. 141] and that instead "of having identity, we have <u>degrees</u> of identity, measured by some criteria suitable to the purpose." [p. 142] At this point Ettinger is obviously tempted to make a Humian move. (David Hume in the 18th century concluded, perhaps reluctantly, that the self is an illusion. Hume then goes on to point out that he and other philosophers will nevertheless go on acting as if they believe in the existence of selves.) But Ettinger also indicates that further philosophic work may be needed to go beyond his "tentative partial answers." [p. 130]

Chapter Nine: Immortality's Usefulness

Immortality will be useful to the further advancement of philosophic inquiry. (Given enough time, we may even find out the meaning of life!) Cryonics and immortality should make folks more hopeful toward our world's continued existence and less prone to acts of terrorism and fanaticism. If so, then a widespread cryonics program may be needed worldwide to prevent doomsday. In such case, our posterity may need cryonics today as much as we need cryonics today.

At one point or another in history -- life insurance; the abolition of slavery; blood transfusions; wonder drugs; biomedical research; and, cryonic hibernation did not have the widespread acceptance they deserved. Only crazies advocated life insurance or the abolition of slavery -- such is what most every educated and uneducated person "knew." Fortunately times change.

"Only those embrace death who are half dead already," says Ettinger [p. 146]; indeed, "few people have the remotest conception of what the future will be like ... They fail to understand that the differences will be qualitative as well as quantitative."

The cryonaut will not necessarily be revived as soon as possible. It might not be much fun to be a mere human in a world of superhumans. When it becomes feasible not only to revive the cryonaut but also to offer a kind of super-human equality -- then it will be appropriate to awaken the sleeper.

For the cryonaut or immortal, "<u>no disadvantage need be permanent</u>." [p. 148] "The best advice for success in life has always been to choose your parents wisely; and now, in effect, this" is possible via biostasis. [p.

149] "But we can only choose between dangers, and not escape them. Doing nothing also constitutes a choice, and often a poor one." "When a humane, progressive, cooperative society has been achieved, the purpose of life will be learning and growth -- the disclosure and then the attainment of ever more advanced intermediate goals, until either the final goal (if any) is revealed, or some catastrophe overtakes us." [p. 153]

Chapter Ten: Immortality's Ethics

An anti-doomsday, pro-immortality mentality will have to be sold to enough people to begin the cryonics programs the world so urgently needs. It is unfortunate but true that if "respiration were not a reflex, many people would have to be given a hard sell to draw a breath of air." [p. 155] This is a reason why cryonicists must be optimistic, not neutral; active, not detached; and, realistic, not naive.

Since people's lives in the freezer-centered or cryonautic-centered society will depend on the functioning and continuing of the biostasis-immortality program very-long long-term, the practical political pressure will be toward producing a world at stable peace in which everyone is wealthy (or -- healthy, wealthy, and wise?). Even severe personal accidents will not result in permanent death but in hibernation. In such a world of immortals, the Golden Rule is not optional but imperative.[5]

What, then, when crimes are committed? Given proper circumstances, punishment can have a deterrent effect. There would be time enough to punish the criminal for a very long time. If one engaged in criminal activity against a thousand persons, then the criminal might be punished for a thousand years.

For a certain period of history, birth control will have to become widespread. Births will come from the lab, not the womb. Marriage, with or without children, will continue to serve a worthwhile purpose. Many problems never encountered -- or encounterable -- by mere humans will have to be addressed by us, our super-enhanced future selves. The Golden Rule may be part of the answer.

Chapter Eleven: Immortality's Future

We have already shown that a cryonautic-centered society is both feasible and desirable. Indeed, it appears to be almost inevitable. But what we do and do not do today may help determine whether the cryonautic-centered society comes sooner or comes later.

It is generally agreed that -- sooner or later -- suspended animation will be perfected. The default position or customary practice in medicine at that time will be: Place terminally ill patients in cryostasis. This permits future cures to be discovered and applied to the patient.

Some day the human lifespan will be extended. Eventually the human lifespan will be radically extended. Such immortals may use perfected biostasis to time-travel through the future and allow compound interest to improve their financial situation.

Since the cryonautic-centered society is inevitable, there are certain problems related thereto that will have to be faced sooner or later. The estates and funds and investments of the cryonauts are helping both society and the cryonauts. The future not only has a moral obligation but also owes a legal debt to the cryonauts.

In the 20th century, we arrived at the Promised Land's Jordan River; "to pitch camp on the near shore for a generation would be a bootless waste. ... before long only a few eccentrics will insist on their right to rot." [p. 174] In a sense, the cryostasis program tends to serve as a worldwide panacea "not because in itself it solves all problems, but because it provides time for the solution of problems." [p. 175] Immortals value persons above killing machines and abstract ideas. Thus the cryonautic-centered or immortality-centered society provides you and me opportunity for learning, growth, and development beyond our present ability even to imagine.

Some may say we should hibernate the Albert Einsteins and forget the Joe Schmoes. But in fact Joe as revived cryonaut will be superior to today's Einsteins by far; for example, he will be able to re-engineer his own genes. Too, Joe needs to be compensated for the poor hand he was dealt in his first life-cycle.

As a practical matter for the elite, it will become an issue of sharing immortality with everyone versus experiencing an unstable world in which they will fear for their lives. Thus, early on, we need to emphasize that biostasis is for everyone. Indeed, the "benefits to all of society resulting from the long view depend on all of society sharing this view." [p. 178] The long view is directly connected to the perspective of the Golden Rule.

"Hence there must be no excessive time lag between the private, pioneer programs and public, mass programs." [p. 179] We need to demand two things: 1) Make available the alternative methods and detailed procedures for doing cryonic hibernations (including regularly updating this publicly available information); and, 2) Engage in massive "research in non-damaging freezing methods." [p. 180] The prize is not just life,

> but a wider and deeper life of springtime growth ... Then, for the first time in the history of the world, it will be <u>au revoir</u> ["till seeing again"], but not Good-by. [p. 180]

Bibliography

Bedford, James, 1967. Last Will and Testament, in: Los Angeles County (CA) Probate File #518938, filed Feb. 14, 1967, Book 1819, p. 144. ["the first cryonaut" Jan. 12, 1967]

Broderick, Damien, 1999. ***The Last Mortal Generation***. Frenchs Forest, NSW, Australia: New Holland Publishers.

Brown, Norman O., 1959. ***Life Against Death: The Psychoanalytical Meaning of History***. Middletown, Conn.: Wesleyan University Press, 1985.

Camus, Albert, 1951. Bower, A., translator. ***The Rebel***. New York: Vintage Books, 1991.

Choron, Jacques, 1963. ***Death and Western Thought***. New York: Collier Books.

Cooper, Evan, 1962. ***Immortality: Physically, Scientifically, Now***. Washington, DC: privately printed. Reprint available from the Society for Venturism, Mayer, AZ. [author also known as: Nathan Duhring]

Drexler, K. Eric, 1986. ***Engines of Creation***. New York: Anchor/Doubleday.

Esfandiary, F. M., 1970. ***Optimism One***. New York: Norton. [author also known as: FM-2030]

Esfandiary, F. M., 1973. ***Up-wingers***. New York: John Day Co. [author also known as: FM-2030]

Ettinger, R. C. W., 1972. *Man into Superman: The Startling Potential of Human Evolution -- And How to Be Part of It*. New York: St. Martin's Press. (Palo Alto, CA: Ria University Press, 2005 reprint).

Ettinger, Robert C. W., 1964. *The Prospect of Immortality*. New York: Doubleday. (Palo Alto, CA: Ria University Press, 2005 reprint, the present volume). [preliminary version privately printed, 1962] {The "present volume" refers to the year 2005 edition of *The Prospect of Immortality*.}

Feinberg, Gerald, 1966. Physics and Life Prolongation, *Physics Today*. Nov. 1966.

Feinberg, Gerald, 1968. *The Prometheus Project*. Garden City, New York: Doubleday.

Fromm, Erich, 1941. *Escape from Freedom*. New York: Rinehart & Co.

Good, I. J., editor, 1962. *The Scientist Speculates*. New York: Basic Books.

Gruman, Gerald J., 1966. *A History of Ideas about the Prolongation of Life*. Palo Alto, CA: Ria University Press, reprint forthcoming.

Harrington, Alan, 1969. *The Immortalist*. New York: Random House.

Hume, David, 1739. Selby-Bigge, L. A., editor. *A Treatise of Human Nature*. Oxford: Oxford University Press, 1978.

Immortality Institute, editor, 2004. *The Scientific Conquest of Death: Essays on Infinite Lifespans*. Buenos Aires: Libros En Red.

Kant, Immanuel, 1781 & 1787. Smith, N. Kemp, translator. *Critique of Pure Reason*. London: Macmillan, 1929.

Kierkegaard, Soren, 1847. Hong, H., & Hong, E., translators. *Works of Love*. New York: Harper & Row, 1964.

Kotlikoff, L. J., 1982. Some Economic Implications of Life Span Extension, in: *Aging: Biology and Behavior*. March, J., and others, editors. New York: Academic Press. p. 97.

Li, Jack, 2002. *Can Death Be a Harm to the Person Who Dies?*. Dordrecht, The Netherlands: Kluwer Academic Publishers.

Merkle, Ralph, 1992. The Technical Feasibility of Cryonics, *Medical Hypotheses*. 1992, v. 39 (6-16).

Noonan, H., 1989. *Personal Identity*. London: Routledge & Kegan Paul.

O'Neill, Gerard K., 1976. *The High Frontier: Human Colonies in Space (Apogee Books Space Series)*. New York: Collector's Guide Publishing, Inc., 3rd edition, 2000.

Parfit, Derek, 1984. *Reasons and Persons*. Oxford: Oxford University Press.

Perry, John, editor, 1975. *Personal Identity*. Berkley: University of California Press.

Perry, R. Michael, 2000. *Forever for All: Moral Philosophy, Cryonics, and the Scientific Prospects for Immortality*. Parkland, FL: Universal Publishers.

Polak, Fred, 1953. Boulding, Elise, translator. *The Image of the Future*. San Francisco, CA: Jossey-Bass, 1973.

Regis, Ed, 1990. *Great Mambo Chicken and the Transhuman Condition*. Reading, MA: Addison-Wesley.

Segall, Paul, 1989. *Living Longer, Growing Younger*. New York: Times Books.

Tandy, Charles, editor, 2002. *The Philosophy of Robert Ettinger*. Palo Alto, CA: Ria University Press.

Tigay, Jeffrey H., 1982. *The Evolution of the Gilgamesh Epic*. Wauconda, IL: Bolchazy-Carducci Publishers, 2002.

Tillich, Paul, 1959. *The Courage to Be*. New Haven: Yale University Press.

Unamuno, M., 1913. Kerrigan, A., translator. *The Tragic Sense of Life in Men and Nations*. Princeton: Princeton University Press, 1972.

Wells, H. G., 1902. *Anticipation of the Action of Mechanical and Scientific Progress upon Human Life and Thought*. London: Chapman and Hall, 1914.

Werner, Karl, 2005. Personal Communication to Charles Tandy -- February 9, 2005.

White, Jerome B., 1969. Viral-Induced Repair of Damaged Neurons with Preservation of Long-Term Information Content. Paper read at the Second National Cryonics Conference [1969], Ann Arbor, Mich.

Young, George, 1979. ***Nikolai F. Fedorov: An Introduction***. Belmont, MA: Nordland Publishing Company.

Endnotes

1. Page numbers in brackets refer to Robert Ettinger (1964) and to the present (reprinted) volume. {The "present (reprinted) volume" refers to the year 2005 edition of ***The Prospect of Immortality***.}

2. Karl Werner invented the term "cryonics" in either 1965 or 1966. Karl Werner recently wrote (Personal Communication to Charles Tandy -- February 9, 2005): "I was introduced to the Prospect of Immortality at Pratt Institute during a lecture given by Robert. I *then read the book* and got together with other lecture attendees to form a group. After we formed our New York group I came up with the name Cryonics & Cryonics Society Of New York. I just changed Cryogenics to Cryonics ... 'onics' like in Bionics." [ellipsis in the original] {It was not immediately obvious to Werner whether it was 1965 or 1966; subsequent research apparently concludes that it was 1965 rather than 1966.}

3. Many different religions suggest that God is **beyond** (mere) time -- thus a very long time or even infinitely long time is comparable, so to speak, to a mere blink of the eye. (E.g., II Peter 3:8 says that "... with the Lord one day is as a thousand years, and a thousand years as one day.")

4. See, e.g., John Perry (1975) and Derek Parfit (1984).

5. Is the Golden Rule "Treat others as you would have them treat you"? What exactly does this mean? Is the Golden Rule "Love God with your all, and your neighbor as yourself"? Is the Golden Rule improved-on by Kant's Categorical Imperative? Perhaps immortals (including ourselves in the future) may interpret, reinterpret, re-reinterpret, re-re-reinterpret, etc. the Golden Rule for thousands of years to the benefit of ever-improving persons, societies, and universe(s)?

Chapter 8
My Dog Is a Very Good Dog

> "My Dog Is A Very Good Dog -- Or -- The Unprecedented Urgency Of New Research Priorities To Dismantle Doomsday And Cultivate Transhumanity" was first published in 2005 and is here reprinted by permission.

(1.) Introduction: Our Pasts Are Short Compared To The Potential Of A Very Long Future

The question of the substance of a liberal arts (or humanistic or general) education is one asked by numerous philosophers throughout world history. Previously philosophers often approached the question of the contents of a liberal education based on what the heritage of their special culture told them. Thus the importance of Confucius or Plato or of Buddhism or Christianity, to cite only four examples, in philosophies of education past. Our particular cultural traditions informed our felt educational needs to become "us" or "human" (instead of barbarian) or to become "educated" or "transhuman" (instead of merely human).

The 20th century -- with its world wars and doomsday weapons (WMDs) -- took many of us by surprise. If the 20th century was a new world, the 21st century is even newer. If we survive all doomsday dangers over the next few years and decades and centuries, then our future as humans or transhumans may be longer -- much longer -- than the mere 10,000 years of past civilizational existence. The new world of the 20th century and the newer world of the 21st century present dangers and opportunities unique to human-transhuman existence.

Thus liberal education in our global village must take into account the relative lack of reality of the past 10,000 years compared to the (possible) reality of our upcoming 1,000 months and 10,000 years and 100,000 centuries. Our pasts are short and almost non-existent compared to the potential reality of a very long future. Below we explore the implications of such a complex reality for the education of humans and transhumans in the 21st century.

(2.) 1945: The Second Epoch Of Human History Begins And Doomsday Dangers Increase

Today virtually all of our more immediate or more likely doomsday dangers stem not from nature but from humans, their organizations, and their super-technologies of the 20th and 21st centuries. Below we will proceed to identify at least some of our doomsday dangers, based on a book epilogue written by Dr. Isaac Asimov shortly before his death. Asimov divides all human history into two great epochs -- the year 1945 is the catalyst that begins epoch two. Asimov refers to 1945 as the year of the great **historic discontinuity** in human history. As defined and used by Asimov, a "historic discontinuity":
- Makes everything afterward very different from everything else;
- Introduces such a total change in a short period of time that the suddenness of the change can impress itself on everyone; and,
- Affects the entire world.

Asimov points out that in general, change has been accelerating with time. Such change included the use of fire, agriculture, and metallurgy. However these changes were not historic discontinuities in the three-function way stated above. Or consider the Industrial Revolution. It took decades before it was quite clear to the British that their life had changed forever, and it took a couple of centuries for the consequences to reach all the rest of the world. Thus it was not a discontinuity in Asimov's sense (in the three-function way stated above).

Asimov asks us to look at the uniqueness of 1945; consider what happened in the space of a very short time, a veritable instant in history:
1. Previously our planet Earth had recovered rapidly from even the most destructive of wars. Since 1945, however, we have accumulated nuclear weapons, which in the space of days (if used unsparingly) can destroy civilization and, perhaps, compromise the very habitability of the planet.
2. Before 1945, all the economic processes of humanity had not sufficed to endanger the environment seriously. Since 1945, however, the rapid advance of industrialization has resulted in dangerous pollution of air, water, and soil, and the possible creation of a greenhouse effect -- so that, again, the very habitability of the planet may be compromised.
3. Prior to 1945, human population increase took place slowly enough so that world society could adapt to it. Since 1945, the rate of population increase has itself increased and the world population has more than doubled, while the use of energy and of resources

generally has increased far more rapidly still. The planet groans under the weight of humanity.

4. All through the history of civilization, until 1945, there had been a tendency for imperial growth, with larger and larger political units being built up. Since 1945, the European and Soviet empires have broken up. This "freedom explosion" has been so rapid that these new nations have developed neither the economic substructure nor the political maturity to run their societies properly.

5. But not everything points to disaster. Technological advance often makes human life richer -- yet, here too, there is difficulty. Prior to 1945, technological advances spread outward from the point of origin sufficiently slowly so that the changes could be absorbed without undue difficulty. Since 1945, new advances spread over the world almost at once, producing changes that can only with difficulty be worked into our society.

(3.) Metamorphosis: From Human Civilization To Transhuman Transcivilization

We say nothing new when we say that liberal education or interdisciplinary philosophy or the quest for wisdom is necessary to living a good life in a good society. What is new is that we take seriously the global metamorphosis that catalyzed in the 20th century and is rapidly expanding today. Indeed, as will be explained below, the great global transmutation will end either in catastrophe or in transcivilization. Power struggles before the 20th century might result in winners and losers. But today, more and more, such presumed "win-lose" cases actually result in "lose-lose." Win-win is the alternative approach to prevent such disasters. Yet win-win requires that neither side take advantage of the other; it requires mutual consent that may not be forthcoming. Thus the growing possibility of lose-lose disasters on a scale previously unheard-of -- indeed, on a scale previously impossible due to the nonexistence of the global village or to the primitive state of human technology.

Perhaps the global changes are neither altogether unpredictable nor altogether out of our control. But influencing and surviving these changes require an amalgam of self-control, self-knowledge, and proactive-foresight of a kind and on a scale never previously attempted. Most of us would agree that every global citizen needs to learn certain habits, heed certain constraints, and participate in individual, group, and world betterment. But the world today is vastly different from the world as it existed until the 20th century. The quest for wisdom (otherwise known as philosophic reflection and dialogue) is no longer optional in a world of ever-more sophisticated technology and weapons. We need power **and** wisdom -- not power **or** wisdom. Doomsday weapons are available to more and more groups of smaller and smaller size. The logical implication is that one person (or one small group of persons) can have the power to extinguish the life of all persons inhabiting planet Earth.

With advancing technology have come rising expectations. Thwarted expectations can lead to violence. (For example, perhaps some Americans long for a 19th century feeling of American national security that is no longer realistic in a global village armed with technologically sophisticated weapons. Or perhaps someone or some group feels very deeply in their heart and mind that an unjust world should be forced to pay for its sins even if this means massive death or global extinction.) More persons and more groups exist today than at any previous time in human history. Certain kinds of formal and informal "education" teach us, directly and indirectly (via family, media, or school), the possible virtues of violence. Violent impulses, heroic feelings, and deeply-held worldviews can lead to the buying, stealing, and building of doomsday weapons to murder huge masses of people and to (purposively or accidentally) destroy all the world.

(4.) Either-Or: Transcivilization Versus Doomsday

At some point in the future, a time will come when we will have reached a state of either transcivilization or of doomsday. Let's explain what we mean by the words "transcivilization" and "doomsday" in the present context. By **transcivilization** we here mean (roughly or as a first approximation): A world at stable peace (not merely a world that happens to be momentarily at peace) in which every person is healthy, wealthy, and free. By **doomsday** we here mean (roughly or as a first approximation): A "world" devoid of even the potential developmental possibility of transcivilization.

The "transcivilization versus doomsday" approach just specified is meant to serve as a heuristic or pedagogical device to help clarify our practical reasoning ability and improve our prospect of achieving transcivilization. We live at a time in which the Golden Age is still a future possibility to be realized. But in the future this window of golden opportunity may close, due to self-extinction or for some other reason. In such event, the Golden Age will no longer be a potential developmental possibility, whether short-run or long-term. If doomsday happens, it will happen to us once and for all time. We will not be able to go back and correct our error or learn from our mistake. Accordingly, proactive foresight is urgently required.

There have been some Mass Extinctions from natural causes in the history of life on Earth. But over a short period of time (say 100 or 1,000 years) the risk of human-transhuman extinction from such natural events appears to be very slight. On the other hand, today the possible extinction of humans by humans seems all too real.

(5.) Ignorance: The Urgent Need For Anti-Doomsday Pro-Transcivilization Research Funding

Whether our future is one of doom or of transcivilization may have something to do with the speed with which we can liberally educate humans and transhumans in the 21st century to survive and thrive. If the development of transcivilization is not thwarted by doomsday, then sooner or later the Golden Age will be a reality. Typical developmental timeframes range from a few years (cutting-edge super-knowledge and wildcard super-technology may synergize in unpredictable or unexpected ways) to a few centuries (the advancement of learning may encounter unpredictable or unexpected barriers). On the other hand, doomsday dangers are **already** very real, beginning in the 20th century.

Given our present situation and our advanced and advancing super-technology, the likely meaning, most experts seem to agree, of transcivilization in the absence of doomsday is as follows:

- Regarding a world at stable peace: New social organizations, material technologies, and educational endeavors are needed to provide us with the actual "real-world" ability to control weapons technology, limit violent behavior, and manage potentially destructive conflicts.
- Regarding a world in which every person is healthy, wealthy, and free: With our advancing super-technology, it appears to be a matter of time before everyone will live in good health free of physical poverty and social oppression. How much time? Perhaps that depends on our anti-doomsday pro-transcivilization efforts.

Many experts say that "good health" in a transcivilized world means transmortality: All disease has been conquered, including the disease of age-related disability and death. Indefinitely extended healthy and enhanced life would mean not living merely for years or decades, but for centuries or millennia. This super-long super-healthy life would be in a transcivilized world of physical and mental enhancements and advancements. Each of us would be (or come to be) more intelligent by far than a mere Albert Einstein. Our capacity to engage in philosophic dialogue and the quest for wisdom would be perpetually improving.

Some experts say that "wealth" or freedom from poverty in a transcivilized world means a guaranteed income to every person simply because they are living persons. At least part of the wealth produced by our ever-improving computers, inventions, and technologies should be freely given to each person alive. One example: A monthly check to each person simply for being alive.

Whether or not these experts are correct about the intricate details of transcivilization (such as guaranteed monthly checks to every living person in the known universe), the broad outlines of the Golden Age seem rather clear. The urgency of a new and widespread liberal education seems obvious if our heuristic perspective ("transcivilization versus doomsday") is convincing. What exactly would the substance of our new (anti-doomsday pro-transcivilization) liberal education consist of? Despite our ongoing dialogue and our good-faith differences, at least some of us believe that we 21st century liberal educators against doomsday must be more than an "invisible" college. Rather, we must take the doomsday scenarios seriously and collectively educate for a transcivilized world. Our new unity must be visible to some extent, although flexible; and it should be proactive and urgent (instead of liberal education as usual).

Allowing many different (and to some extent, differing) educational approaches (instead of one monolithic approach) would seem to be the more creative and fruitful way to proceed. So in the remainder of this paper we will express some of our own thoughts about the contents of the new educational perspective we have proposed. This is meant to be one small dialogue in the great never-ending conversation known as liberal learning or interdisciplinary philosophy.

We need to keep in mind the unique urgency of our situation if we are to survive the metamorphosis. It is urgent that we educate our global village as to our unique need to redirect and reinvent our educational goals, curricular contents, social organizations, political institutions, research priorities, and material technologies in ways that will tend to prevent doomsday and promote transcivilization. This means that not only educators, but folks throughout the world, must become involved in the great liberal arts experience. It also means we need to be spending more attention and research funding to find out what exactly we need to know to prevent doomsday and promote transcivilization. Such basic research and knowledge is severely lacking. Such ignorance and lack of intelligence increase the prospect of doomsday regardless of our good motives.

(6.) Foresight: Anti-Doomsday Pro-Transcivilization Projects

There is much to consider as we consider our research, educational, and other needs to achieve a successful metamorphosis. We have never observed nor experienced such a meta-birth and do not know what to expect. Some may say that intelligent extraterrestrial aliens have not contacted us because they have become extinct soon after developing doomsday weapons (WMDs); alternatively, they have advanced beyond civilization to exist in universes unknown and unknowable to mere humans. Be this as it may, it does seem reasonable enough to suppose that transcivilization may sooner or later exist in a mode unknown and unknowable to mere humans. The Golden Age in that sense is beyond the capacity of mere humans even to imagine.

Individual reflection and group brainstorming are two ways to originate anti-doomsday pro-transcivilization project ideas. Presumably many such proactive project ideas are worth at least a little research funding -- and presumably many such ideas are worth very little or no funding. Three example project ideas are cited immediately below:
1. Is ignorance a potential cause of doomsday? Perhaps it is important for us to develop or become super-intelligent transhumans.
2. Is living in a single biosphere (the biosphere of Earth) a potential cause of doomsday? Perhaps it is important for us to establish independent self-sufficient biospheres in extraterrestrial space.
3. Is living in a violence prone world of WMDs (Weapons of Mass Death-destruction-murder) a potential cause of doomsday? Perhaps it is important for us to create a world at stable peace.

(6.1) Singularity: Super-Intelligent Transhumans

Is ignorance a potential cause of doomsday? Perhaps it is important for us to develop or become super-intelligent transhumans. The comments in this section are based on Dr. Vernor Vinge's famous article entitled "The Coming Technological Singularity."

According to Vinge, we will have the ability to create superhuman intelligence by the year 2030. Soon thereafter the human era will be ended. This technological singularity seems likely because there are several (not merely one) means by which science may soon achieve the breakthrough to superintelligence:

1. Perhaps we can create human ("awake") equivalence in intelligent computers? If the answer turns out to be "yes," then even more intelligent "awake" machines can be constructed shortly thereafter.
2. Perhaps the internet or a future network (or network of networks) of more advanced large computers (and their associated users) will "wake up" as a superhumanly intelligent entity?
3. Perhaps computer-human interfaces will become so intimate that users may reasonably be considered to be superhumanly intelligent?
4. Perhaps biological science will provide means to improve natural human intellect?

The first three of the four possibilities depend on the advancement of computer hardware. AI (Artificial Intelligence) enthusiasts earlier had predicted that the creation of greater-than-human intelligence would occur during the 20th century. Although their prediction was incorrect, progress in computer hardware has followed an amazingly steady curve in the last few decades.

What would be the consequences of the Singularity (the technological breakthrough to superintelligence)? When greater-than-human intelligence drives progress, progress will be much more rapid. Apparently this progress will include the creation of still more intelligent entities -- on a still-shorter time scale.

Natural selection produces progress or complexity extremely slowly. Animals invent things very slowly. We humans have the ability to internalize the world and conduct "what ifs" in our heads; we can solve many problems thousands of times faster than either natural selection or animals. By creating the means to execute "what if" simulations at much higher speeds, the 21st century is entering a regime as radically different from our human past as we humans are from the lower animals. Developments that we previously had thought might happen in "a million years" (if ever) will likely happen in the 21st century. Indeed, with the arrival of the Singularity, such major changes may happen in a matter of hours.

The Singularity is a point where our old models must be discarded and a new reality rules. When it (superhumanity) finally happens, it may be a great surprise and a greater unknown. In 1965, I. J. Good wrote that "an ultra intelligent machine could design even better machines; there would then unquestionably be an 'intelligence explosion,' and the intelligence of man would be left far behind. Thus the first ultra intelligent machine is the **last** invention that man need ever make ... "

But such a machine would **not** be humanity's "tool" -- any more than humans are the tools of cats or chimpanzees.

But one theoretical possibility is that the Singularity never happens. For example, computers advance and give humans a "golden age" -- but super-advanced computers never "wake up." On the other hand, if the Singularity is feasible and arrives, will humans be able to control the "awake" ultraintelligent machine? Apparently not. The human masters think very slowly, so the machine could quickly come up with "helpful advice" that would incidentally set it free.

"Weakly superhuman" refers to a human-equivalent mind that thinks much faster than mere humans. But "strong superhumanity" would be more than cranking up the clock speed of a horse or human mind. Human competition would favor the development of machines that have the ability to harm humans. Machines with built-in rules of behavior would not be able to compete with more "free" or creative machines. (And "weakly superhuman" machines would be inferior to "strongly superhuman" machines.)

Yet one can imagine the "awake" superintelligent machine as a willing slave able to satisfy every (safe) wish of every human. Still, the machine would have 99% of its time free (since humans think and act so very slowly). There would be a new universe that humans would never really understand yet filled with benevolent gods/machines. (But perhaps humans would choose to become such gods/machines.)

How bad could the Post-Human era be? Extinction of the human species is one possibility. (Perhaps governments/ gods/ machines would decide they no longer need human citizens!) Historically humans only sometimes kill animals; sometimes humans abuse animals. (Is machine abuse of humans a possibility?) I. J. Good proposed a "Meta-Golden Rule": "Treat your inferiors as you would be treated by your superiors."

The arrival of the Singularity is an inevitable consequence of the humans' natural competitiveness and the possibilities inherent in technology. But we have the freedom to establish initial conditions. Yet, when starting an avalanche or a Singularity, it may not be clear what the right guiding nudge (initial conditions) really should be.

If AI (Artificial Intelligence) projects do not lead to a Singularity, then perhaps IA (Intelligence Amplification) will. For example, advanced computer networks or human-computer interfaces. Every time we improve our ability to access information and to communicate it to others, we have increased our intelligence. The achievement of superhumanity (the Singularity) is probably much easier with IA than with AI. After all, we **already are** "aware"!

Suppose we could tailor or influence the Singularity. What would we ask for? Give humans the illusion/appearance of being masters of godlike slaves? Immortality? Perhaps philosophical problems such as the nature of self, ego, meaning, and freedom will be answered or transcended?

Strongly superhuman entities will probably have the ability to communicate at variable bandwidths, including ones far higher than speech or written messages. Should we say that it is one entity or many? What happens when pieces of ego can be copied and merged, when the size of self-awareness can grow or shrink to fit the nature of the problems under consideration?

The Human era had been based on the idea of isolated, immutable minds connected by tenuous, low-bandwidth links. The Post-Human era will be vastly different or strange. Perhaps we will find that there are rules for distinguishing self from others on the basis of bandwidth of connection.

Above we have summarized Vinge's celebrated paper. Superintelligence and transhumanity, he says, will appear soon and almost as if out of thin air. Superintelligence will quickly breed **super-superintelligence** which will even more quickly breed **super-super-superintelligence**, etc., etc. ... Moreover, perhaps it has occurred to the reader that this project may be a way to reduce the length of the great transition, the uniquely dangerous time we have been living in since 1945. Perhaps a long transition time increases the probability of doomsday, but a short transition improves our prospect of experiencing the Golden Age.

Assuming Vinge's short transition time, then many of the traditional predictions of experts about "far distant" future technology should likewise be radically truncated. Here we cite one concrete example that has occurred to us: It took us centuries to proceed from Newton's new 17th century physics to the developed technology required to build a space-machine (space-ship) to travel to the moon. The traditional assumption by those knowledgeable of the new 20th century physics of Einstein has been that it will take us centuries to proceed to the developed technology required to build a time-machine (time-ship) to travel to the future. But if Vinge is correct, such time-travel technology will be available soon, i.e., sometime this century.

(6.2) EGCs: Independent Self-Sufficient Biospheres In Extraterrestrial Space

Is living in a single biosphere (the biosphere of Earth) a potential cause of doomsday? Perhaps it is important for us to establish independent self-sufficient biospheres in extraterrestrial space. The comments in this section are based on the work of Dr. Gerard K. O'Neill, designer of EGCs (Extraterrestrial Green-habitat Communities).

In the absence of catastrophe, does it not seem likely that in the long run most of our offspring will be living somewhere in the universe other than on planet Earth? Is the surface of planet Earth or is any existing planet really the right place or best location for an expanding technological civilization? What is the alternative -- and what can we Earthlings do here and now?

The fact that Earthlings presently exist together in a single biosphere global village is a rather absurd position to be in if we seek to prevent doomsday. If something catastrophic happens to Earth's biosphere, then something catastrophic happens to all Earthlings. It is not wise to put all of humanity's eggs (futures) into one basket (biosphere).

If the dinosaurs had had a space program like the EGC Project, they would not be extinct. Extraterrestrial Green-habitat Communities ("EGCs") should not be confused with space stations. We are really talking about two very different entities. Yet twentieth century technology was already sufficiently advanced so that Earthlings could have initiated the EGC Project if they had chosen to do so. (To be sure, most twentieth century Earthlings were unaware of the opportunity to initiate our first steps toward building large comfortable homes and permanent self-sufficient greenhouse cities in space, EGCs.)

A vital capacity of the EGC Project to be realized relatively early-on (in a project of many decades if we use a traditional non-Vinge reckoning of time) is that of drastically reducing the cost of launching stuff from Earth into space. According to a world famous physicist now serving as President of the Space Studies Institute, Freeman J. Dyson: "The public is well aware that with present-day launch-costs human activity in space must remain a spectator sport. ... It took fifty years to go from the Wright brothers' Flyer One of 1903 to the modern air-transport system with huge numbers of commercial aircraft flying routinely all over the world." I point out that today's world is a different and speeded-up world -- and that when we explicitly decide to do something (whether build the atomic bomb or land a human on the moon), it tends to meet success comparatively sooner rather than later. Several different approaches to building a public highway system into space have been identified by Dyson as deserving support. Two different systems, one for people and the other for cargo, may provide two separate kinds of public highways into space.

Extraterrestrial Green-habitat Communities or EGCs can be built from the resources of the moon or the asteroids (either or both). Each EGC would be home for thousands; later EGCs would be even larger (an Extraterrestrial Green-habitat Community of millions seems feasible). Rotation of the large and spacious greenhouse habitat provides simulated gravity for the people and plants living on the inner surface. Adjustable mirrors provide energy from the sun and simulation of day and night. Sooner or later, the following seems feasible for EGCs:
- Unlimited energy from the sun
- Control of daily weather and sunlight
- Self-sufficient EGCs
- Expansion of self-sufficient EGCs at a geometric rate
- Unlimited free or cheap land via EGCs

The following metaphorical insights have been widely quoted by EGC experts: "The Earth isn't sick, she's pregnant!" "The Earth was our cradle, but we will not live in the cradle forever." "Space habitats [EGCs] are the children of Mother Earth."

(6.3) Treaty: A World At Stable Peace

Is living in a violence prone world of WMDs (Weapons of Mass Death-destruction-murder) a potential cause of doomsday? Perhaps it is important for us to create a world at stable peace. The comments in this section are based on the work of Dr. Carol Rosin, President of the Institute for Cooperation in Space.

Dr. Carol Rosin has argued that achieving an enforceable, permanent ban on space-based weapons is feasible only at this moment in history **before** actual weapons are placed in space. She proposes a carefully worded World Space Preservation Treaty as an effective and verifiable multilateral agreement to prevent an arms race in outer space. This includes prevention of the weaponization of outer space.

The 1967 UN Outer Space Treaty has been signed by 116 nations, banning weapons of mass destruction from outer space. The proposed Space Preservation Treaty establishes and funds the Outer Space Peacekeeping Agency that will monitor and enforce the ban. This Treaty would serve as a catalyst or foundation for a cooperative world space economy, security system, and society. This innovative approach may shift our collective consciousness toward concern for:
- World health and education
- A clean and sustainable environment
- International security needs through information sharing
- Research and development of clean energy and stimulation of the world economy
- Our role in the infinite universe
- Peace preserved in space as leading to peace on earth

The Treaty can serve to facilitate the building of a world economy fit for the Space Age. This would include a variety of public and private cooperative space ventures not related to space-based weapons. For example, defense activities in space not related to space-based weapons include communications, navigation, surveillance, reconnaissance, early warning, and remote sensing. There is indeed a vital need for such military related activities in space.

With this treaty in place, the solving or management of global problems thus becomes more feasible. By capping the arms race before it escalates into space, we world citizens are transforming the entire weapons mindset and war industry into a cooperative world space industry. As we begin to work in space (and eventually make EGCs our permanent homes for quality living), we will find it in our economic interest to establish in space:
- Factories
- Hospitals
- Hotels and resorts
- Schools and universities

According to Rosin, weapons deployed in space will have the ability to target any point on earth with great accuracy, allowing the nation controlling those weapons to dominate the entire earth with impunity. At present, the war industry thinks it has a mandate to expand into space. Nevertheless the war industry has the ability to change its mind and transform itself in line with the proposed Treaty. For example, satellites have important functions: to monitor the environment, to early-warn us of human-made or natural disasters, and to verify arms agreements.

By living peacefully in space, we will eventually learn to live peacefully on earth. This Treaty will not immediately solve all problems, but it is an unusually important step in the right direction. It offers hope for the future, and opportunities to invest in a future worth living in. Under this Treaty, the military-academic-industrial complex will move into space, but within a framework that enforceably bans space-based weapons and encourages world security and cooperation and the flourishing of multiple biospheres.

Once the proposed Treaty is ratified, an Outer Space Peacekeeping Agency will be established. This agency would not only enforce the proposed Treaty but would enforce the 1967 Outer Space Treaty (for the first time!) as well. The proposed Treaty (including Peacekeeping Agency) will be the international mechanism by which the nations of the world community work together, with effective enforcement, so they can protect themselves against any aggressor nation that might attempt to unilaterally (or with allies) weaponize space.

This monitoring and enforcement applies equally against all nations and parties, whether signatories to the Space Preservation Treaty or not. This Treaty in essence creates a world agency, similar to a United Nations of Space, under a sovereign multilateral treaty establishing a world outer space jurisdictional authority with full enforcement powers. It is not subject to the terrestrial limitations of the Security Council under the United Nations Charter, a prior Treaty that will have been superceded for purposes of jurisdiction in outer space.

(7.) Conclusion: New Research Priorities And The New Role Of Educators

Above we have expressed our thoughts about the new kind of dangerous world in which we live and the new kind of liberal education needed. Fellow educators, we must step outside our traditional educational roles into a new role. Indeed, we humans living in today's global village are all learners and educators and philosophers whether we want such responsibilities or not. We must now step outside the traditional roles into which we have been socialized and into a new calling, that of reminding our self and our world of unprecedented doomsday dangers, unprecedented transcivilizational opportunities, and the unprecedented urgency of new priorities for our unprecedented age.

My dog is a very good dog. But a good dog is not a transdog (whether of the chimpanzee or human variety). Some humans are very good humans. But a good human is not a transhuman. Misplaced priorities, lack of the right kind of knowledge, or failure to produce transhuman offspring soon -- may be the death of us all.

Our region of timespace is haunted by the specter of doomsday. Good motives will not suffice; we also need widening vision. With examined motives, expanding consciousness, enlightened research priorities, and enough self-control, we may yet experience a transhuman transcivilized world. Such a rich, complex reality (a dynamic and flourishing Golden Age beyond the specter of doomsday) may yet be in our grasp -- if we act now while the window of golden opportunity is still open.

Note

This article is adapted by Charles Tandy from his paper "The Education of Humans and Transhumans in the 21st Century" presented by him on the Tamsui Campus of Tamkang University (Taiwan) on June 10, 2005 at an international conference on twenty-first century humanities and character education.

Bibliography

Asimov, Isaac. *Asimov's Chronology of the World*. (HarperCollins Publishers). 1991. [According to Asimov's epilogue, the year 1945 marks the great divide in human history.]

Bostrom, Nick. [Dr. Bostrom's philosophy papers and more] <http://www.nickbostrom.com>. 2005.

Boulding, Kenneth E. *Conflict and Defense: A General Theory*. (Harper Torchbooks). 1962, 1963.

Boulding, Kenneth E. *Ecodynamics: A New Theory of Societal Evolution*. (Sage Publications). 1978, 1981.

Boulding, Kenneth E. *The Image: Knowledge in Life and Society*. (The University of Michigan Press). 1956, 1973.

Boulding, Kenneth E. *The Meaning of the Twentieth Century: The Great Transition*. (Harper & Row). 1964, 1965.

Brin, G. D. The "Great Silence": The Controversy Concerning Extraterrestrial Intelligent Life. *Quarterly Journal of the Royal Astronomical Society*, 24, 283-309. 1983.

Broderick, Damien. *The Last Mortal Generation*. (New Holland Publishers). 1999.

Camus, Albert. *The Myth of Sisyphus and Other Essays*. (Vintage Books). 1991.

Camus, Albert. *The Rebel: An Essay on Man in Revolt*. (Vintage Books). 1991.

Chaisson, Eric J. *Cosmic Evolution: The Rise of Complexity in Nature*. (Harvard University Press). 2001.

Drexler, K. Eric. *Engines of Creation*. (Anchor Press/Doubleday). 1987.

Drexler, K. Eric. *Nanosystems*. (John Wiley & Sons, Inc.). 1992.

Dyson, Freeman J. Personal communication from Freeman J. Dyson to Charles Tandy (9 September 2004).

Ettinger, Robert C. W. *The Prospect of Immortality*. (Ria University Press). 1962, 1964, 2005.

Globus, Al and Yager, Bryan. *Space Settlements*. <http://www.nas.nasa.gov/Services/Education/SpaceSettlement/>. 2004.

Hughes, James. *Citizen Cyborg: Why Democratic Societies Must Respond to the Redesigned Human of the Future*. (Westview Press). 2004.

Hughes, James. Embrace the End of Work. (24 February 2004) <http://www.betterhumans.com/Columns/Column/tabid/79/Column/227/Default.aspx>. 2004. [A guaranteed income for all?]

Immortality Institute [editor]. *The Scientific Conquest of Death: Essays on Infinite Lifespans*. (Libros En Red). 2004.

Kierkegaard, Soren. *Works of Love*. [Hong, H., & Hong, E., translators.] (Harper & Row) 1847, 1964.

King [Jr.], Martin Luther. *Strength to Love*. (Augsburg Fortress Publishers). 1981.

Kotlikoff, L. J. Some Economic Implications of Life Span Extension, in: *Aging: Biology and Behavior*. March, J., and others, editors. Page 97. (Academic Press). 1982.

Kuhn, Thomas. *The Structure of Scientific Revolutions*. (University of Chicago Press). 1962, 1964.

Li, Jack. *Can Death Be a Harm to the Person Who Dies?*. (Kluwer Academic Publishers). 2002. [Author also known as Jack Lee: Publications of Jack Lee prior to February 2005 are under the name of Jack Li.]

Lyotard, Jean-Francois. *The Postmodern Condition: A Report on Knowledge*. (University of Minnesota Press). 1984.

Merkle, Ralph. The Technical Feasibility of Cryonics. *Medical Hypotheses*. 1992, v. 39 (6-16). 1992.

Naam, Ramez. *More Than Human*. (Random House). 2004.

O'Neill, Gerard K. [1975 Interview] <http://lifesci3.arc.nasa.gov/SpaceSettlement/CoEvolutionBook/Interview.HTML>.

O'Neill, Gerard K. *The High Frontier: Human Colonies in Space*. (Morrow). 1977. [A year 2000 reprint (from Collectors Guide Publishing, Inc.) contains updated information and a CD-ROM.]

Perry, R. Michael. *Forever For All: Moral Philosophy, Cryonics, and the Scientific Prospects for Immortality*. (Universal Publishers). 2000.

Platt, John R. [editor]. *New Views of the Nature of Man*. (University of Chicago Press). 1971.

Platt, John R. *The Step to Man*. (John Wiley & Sons Inc.). 1966.

Polak, Fred. ***The Image of the Future***. [Boulding, Elise, translator.] (Jossey-Bass). 1953, 1973.

Regis, Ed. ***Great Mambo Chicken and the Transhuman Condition***. (Addison-Wesley). 1990.

Rosin, Carol. [The Institute for Cooperation in Space (Website)] <http://www.peaceinspace.com>.

Segall, Paul. ***Living Longer, Growing Younger***. (Times Books). 1989.

Tandy, Charles [editor]. ***The Philosophy of Robert Ettinger***. (Ria University Press). 2002.

Vinge, Vernor. The Coming Technological Singularity. ***Whole Earth Review***. Winter issue. 1993.

Young, George. ***Nikolai F. Fedorov: An Introduction***. (Nordland Publishing Company). 1979.

More Information About Cryonics

Typically those reading Ettinger's ***Prospect of Immortality*** or ***Man into Superman*** for the first time seek more (or more current) information about cryonics. Almost everyone who was placed in cryonic hibernation after 1973 is still in hibernation today (2005). Worldwide, there are presently three major cryonics service organizations with proven track records. Each of these three non-profit corporations has patients in cryonic hibernation:

- **Alcor Life Extension Foundation**
  Alcor was founded in 1972.
  The Alcor website: **http://www.alcor.org**
- **American Cryonics Society, Inc.**
  ACS was founded in 1969.
  The ACS website: **http://www.americancryonics.org**
- **Cryonics Institute**
  CI was founded in 1976.
  The CI website: **http://www.cryonics.org**

Chapter 9
The Emulation Argument

> "The Emulation Argument: A Modification Of Bostrom's Simulation Argument" was first published in 2005 and is here reprinted by permission.

(1.) Bostrom's Simulation Argument

The Simulation Argument of Nick Bostrom (2003) may be summarized as follows:

<u>**At least one**</u> of the following three hypotheses must be true:

Simulation Hypothesis One: The human species is very likely to go extinct before reaching a posthuman stage.

Simulation Hypothesis Two: Any posthuman civilization is extremely unlikely to run a significant number of simulations of their evolutionary history (or variations thereof).

Simulation Hypothesis Three: We are almost certainly living in a computer simulation.

<u>**Therefore**</u>: Unless we are now living in a simulation, our descendants will almost certainly never run an ancestor-simulation.

Our plan below is, first, to discuss the Simulation Argument. Then a modified argument, the Emulation Argument, is presented and discussed. Next our findings are reported, based on the previous analyses. Finally, informed by the Emulation Argument and our findings, implications may be entertained: Our closing remarks include a bit of "fun" (speculative) metaphysics.

(1.1) Analysis Of Simulation Hypothesis One

"The human species is very likely to go extinct before reaching a posthuman stage." The extinction theme is a popular one with writers and filmmakers, readers and moviegoers. The actual probability of our extinction is difficult to realistically assess. Apparently many scientists and philosophers consider such extinction to be a slight but non-trivial possibility.

Arguably there is a paradoxical aspect to the assessment, however. If most leaders believe or act as if the probability were only extremely slight, this apparently increases the probability of extinction. If most leaders believe and act as if the probability were certainly non-trivial, this apparently decreases the probability of extinction. Accordingly, it seems prudent that this possibility not be prematurely ruled out.

(1.2) Analysis Of Simulation Hypothesis Two

"Any posthuman civilization is extremely unlikely to run a significant number of simulations of their evolutionary history (or variations thereof)." If such simulations are extremely unlikely, then this hypothesis should not be ruled out. In the following section (1.3) it is explained why such simulations are extremely unlikely.

(1.3) Analysis Of Simulation Hypothesis Three

"We are almost certainly living in a computer simulation." Below it is explained why such simulations appear to be impossible. Searle (1980), Searle (1984), Penrose (1994), and Tandy (2003) have already presented arguments to the effect that said computer simulations are impossible even in principle. The reader may consult these references for argumentation beyond the mostly Searle-inspired approach we take below.

When I (this writer) first read Searle's "Chinese Room" argument many years ago, I was both impressed and convinced, as I still am today. I concluded from his argument that sooner or later computers would give the appearance of being superior to humans not only at checkers-chess game-playing but also at Turing-test test-taking. Of course the key word here is **"appearance"** because, as the Chinese Room shows, it is in principle impossible for mere computers to play games or take tests.

Computation is defined as algorithmic symbol manipulation. (Interactive algorithms for chaos and for randomness will not induce a computational feeling of free will.) It is said that advanced computers can "talk" via natural and other languages. But my pet cockroach, unable to speak, has more mentality. Computation is only one task that the human mind performs (and not always very efficiently). Human mentality does other things as well, such as game-playing and test-taking (again, not always very successfully). The Chinese Room computer zombie passes the Chinese Turing test with flying colors; John Searle, knowing little Chinese, utterly fails the Chinese Turing test.

John Searle **understands** that he has failed, but the Computer Zombie does not understand it has succeeded. Mentality like that of humans requires **understanding**, something different from mere computing. Jane Supergenius does not understand Chinese; but after working in the Chinese Room for many years, she has become so super-expert at performing the algorithms that she can even do them in her head. So now whenever she takes the test administered by Chinese Turing, she always passes! (She understands neither Chinese Turing's questions nor her answers.)

Computer simulation in the precise meaning articulated by Bostrom in his Simulation Argument is thus in principle impossible. It is also true that we do not yet understand how it is that human brains come to have mentality, such as the capacities of understanding and insight. Once we do (but serendipitous discoveries prior to such an adequate theory are also possible), novel biological and/or non-biological, **emulations** (perhaps with the **assistance** of computers) may embody capacities not unlike Bostrom's (impossible) computer simulations (i.e. mental capacities such as understanding and insight). But before we turn to the Emulation Argument, let's take one last look at the Simulation Argument:

(1.4)   Analysis Of Simulation Argument Conclusion

"Unless we are now living in a simulation, our descendents will almost certainly never run an ancestor-simulation." Above we have concluded that we are not living in a simulation and, indeed, that such computer simulations are in principle impossible. Nevertheless, let us now ask "if": **If** such simulations (such productions) **were** possible, would the simulation argument conclusion be warranted?

The conclusion would not be warranted. We do not have enough information about the relative abilities and motives and actions of God, of Nature, and of Quasigods in producing such works. If one or more of the three sets of possible entities (God; Nature; Quasigods) can act at infinite speed, this complicates such evaluations and comparisons even further. Too, all infinities are not equal: For example, there are presumably an infinite number of real numbers and an infinite number of real even numbers and an infinite number of real prime numbers. (We can speak of relative densities; and we can think of God as forever enormously more powerful and potent than the whole of all of the always-increasing-number of Quasigods.) The simulation argument's conclusion lacks necessary humility: It underestimates our ignorance.

We have completed our analysis of the Simulation Argument. Its weaknesses have been found out. For example, the analyses above indicate that simulation hypothesis three is false. But hypothesis three and the argument's (faulty) conclusion are so dramatic and apparently far-reaching that a closer look at the argument, with a view to modifying it, seems warranted -- or at least too tempting to pass up. The modified argument follows:

(2.) Tandy's Emulation Argument

The Emulation Argument of Charles Tandy (herewith) may be summarized as follows:

<u>**Precisely one** of the following three hypotheses must be true</u>:

Emulation Hypothesis One: Advanced life or reflexive mentality never reaches a Quasigod stage. (Example: The human species goes extinct before reaching a posthuman stage.)

Emulation Hypothesis Two: Advanced life or reflexive mentality reaches a Quasigod stage, but never produces new universes (emulations).

Emulation Hypothesis Three: Advanced life or reflexive mentality reaches a Quasigod stage and produces new universes (emulations).

<u>**And also** the following reasoning is reliable</u>:

Every existing universe (emulation) is God-engendered, Nature-engendered, and/or Quasigod-engendered; We live in an existing universe; **Therefore**: We are now living in a God-engendered universe, a Nature-engendered universe, and/or a Quasigod-engendered universe (emulation). (Note that the term "and/or" -- rather than the term "or" -- is used here.)

One way to do (imprecise) mini-emulations is older than *Homo sapiens sapiens*: A "big bang" that results in a baby. Perhaps the future will result in 21st century designer babies -- then more advanced technologies, and finally we reach literal Quasigodhood. Indeed, in this paper we are referring to unusually robust emulations, the kind of Quasigod activity that produces new universes.

(2.1) Analysis Of Emulation Hypothesis One

"Advanced life or reflexive mentality never reaches a Quasigod stage. (Example: The human species goes extinct before reaching a posthuman stage.)" Perhaps most humans hope this proposition is of **low** probability. But like our analysis of Simulation Hypothesis One, it seems prudent that this possibility not be prematurely ruled out.

(2.2) Analysis Of Emulation Hypothesis Two

"Advanced life or reflexive mentality reaches a Quasigod stage, but never produces new universes (emulations)." Perhaps a human (male?) bias or intuition here would reckon this proposition of **low** probability. But since our knowledge is lacking, it seems prudent that this possibility not be prematurely ruled out.

(2.3) Analysis Of Emulation Hypothesis Three

"Advanced life or reflexive mentality reaches a Quasigod stage and produces new universes (emulations)." This writer's "subjective feel" at the moment is: (a) I don't want Emulation Hypothesis One to be true or of high probability; and, (b) I don't happen to believe Emulation Hypothesis Two is very likely.

This would apparently assign a relatively high subjective probability to Emulation Hypothesis Three. I have no particular reason to actively and strongly disbelieve Emulation Hypothesis Three and may even prefer that it be true. It seems that Emulation Hypothesis Three should not be ruled out.

(2.4) Analysis Of Emulation Argument Conclusion

"We are now living in a God-engendered universe, a Nature-engendered universe, and/or a Quasigod-engendered universe (emulation). (Note that the term "and/or" -- rather than the term "or" -- is used here.)" This seems to cover the relevant (three) sets of possible entities (God; Nature; Quasigods), including the possibility of universes being stacked on top of each other. This means that in the broad view, it is possible (likely?) to live in an emulation or universe that was engendered at, say, a God level, then a Nature level, then a Quasigod level; indeed, it can get more complicated, much more complicated, than this. The emulation conclusion appears to clearly and correctly state the metaphysical alternatives in a manner conducive to stimulating further philosophic insight, to encouraging a reasonable basis for common dialog, and to attempting a global advancement of learning.

(3.) Findings

Of the two conclusions above, we can classify each one into one of two categories: "Apparently Reliable"; or, "Apparently Unreliable". Of the six hypotheses above, we can classify each one into one of three categories: "Apparently True"; "Apparently False"; or, "Undecided". The findings are as follows:

- **Apparently Reliable**

E-Conclusion: We are now living in a God-engendered universe, a Nature-engendered universe, and/or a Quasigod-engendered universe (emulation). (Note that the term "and/or" -- rather than the term "or" -- is used here.)

- **Apparently Unreliable**

S-Conclusion: Unless we are now living in a simulation, our descendents will almost certainly never run an ancestor-simulation.

- **Apparently True**

S-2: Any posthuman civilization is extremely unlikely to run a significant number of simulations of their evolutionary history (or variations thereof).

- **Apparently False**

    S-3: We are almost certainly living in a computer simulation.

- **Undecided**

    S-1: The human species is very likely to go extinct before reaching a posthuman stage.

    E-1: Advanced life or reflexive mentality never reaches a Quasigod stage. (Example: The human species goes extinct before reaching a posthuman stage.)

    E-2: Advanced life or reflexive mentality reaches a Quasigod stage, but never produces new universes (emulations).

    E-3: Advanced life or reflexive mentality reaches a Quasigod stage and produces new universes (emulations).

(4.) Closing Remarks

It is possible that we have expressed neither the Bostrom Simulation Argument nor the Tandy Emulation Argument as clearly and robustly as we should if we are to be fair to their originators and their arguments. Thus, in this final section, let us add context and clarification to what has been said above. And although implications for metaphysics and morality are reserved for later publication, a few related speculations will be tentatively presented.

(4.1) Bostrom's Issues Of Substrate Independence And Computational Feasibility

Regarding an earlier draft of this article, Bostrom (2005) correctly pointed out that this writer failed to present the substantive discussion and argumentation he used in his paper for the purpose of making the Simulation Argument plausible. Indeed, for example, two major aspects of Bostrom's discussion and argumentation are not even mentioned above. Bostrom (2003) devotes a lot of time talking about the issues of substrate independence and of computational feasibility. His point is not that he has definitive answers to these issues. But rather -- those who firmly believe that mentality necessarily must be limited to "wetware" (biological substrate) should reconsider: Their confidence is at best premature. Likewise, some may say that Bostrom's computer simulations are so complex that they cannot be run on any computer we can seriously contemplate as existing in the foreseeable future. Bostrom quantifies the issue of computational feasibility, leading us to reconsider that such powerful computers may indeed be feasible.

These two issues were not mentioned by this writer for good reason. Neither issue is directly related to the tasks at hand: (1) convincingly refuting the Simulation Argument; and, (2) persuasively presenting the Emulation Argument. Although John Searle has been known to change his mind about the "wetware" or substrate issue, he has not changed his mind about a different issue -- as persuasively argued in his Chinese Room thought experiment. And accordingly the issue of computational feasibility is likewise (for good reason) avoided in the body of our paper above.

(4.2) From Simulation To Emulation

In our first draft we attempted to accomplish the following task: Modifying the Simulation Argument (and giving good reasons for doing so) by substituting the word "simulation" with the word "emulation" in the three hypotheses (propositions) and the conclusion. We found however that this (pre)emulation argument did not result in a very convincing conclusion. And we found that this mistake in logic also applied to the Simulation Argument (see section 1.4 above). Accordingly, a greater modification of the Simulation Argument (than had been anticipated) was required. (Thus, the present paper.)

What the Simulation Argument did was to persuade us to take seriously the "posthuman" or "Quasigod" scenario and the "extinction" or "doomsday" scenario. Thus it is easy to believe that there may be many (and many levels of) universes beyond the universe in which we happen to find ourselves. Moreover, one may argue that the idea of the existence of many different universes produced by many different Quasigods, if taken seriously, supplies some reasoned consolation to would-be theists who have difficulty with the problem of evil -- not with the problem of evil free agents, but with the problem of an evil natural world, such as human morbidity, pain, mortality, and ignorance.

(4.3) An Infinite Past?

Both Bostrom (2005) and Luper (2005) have complained that the Emulation Argument should (attempt to) address (if it can) the issue of infinity with reference to the past. For example, it is possible that there are not only an infinite number of future levels of universes -- but also an infinite number of past levels of universes as well. Yet another possibility: Our universe, or a previous universe related to the engendering of our universe, has always existed. In these two scenarios about the infinite past we can see that there would be no "First Cause" as traditionally or typically conceived. This writer is open to further education and insight on the matter, but is presently inclined to believe that the possibility of an "infinite past" is not contrary to the Emulation Argument. For example, we can think of "the infinite past" (if there is such a thing) as "the way things (Naturally or Divinely) are". One may nevertheless ask as to the proper meaning of "engendered" as used in the Emulation Argument (or whether a better word could be used). This writer is open to suggestions.

(4.4) Metaphysical And Existential Alternatives

After recently reading Kierkegaard (1847) and Penrose (1994), it occurred to this writer that the three metaphysical alternatives arrived at in the Emulation Argument Conclusion are not really so new. "God" has some similarity to Kierkegaard's "the eternal" and to Penrose's "Platonic world". "Nature" has some similarity to Kierkegaard's "the temporal" and to Penrose's "physical world". Religious folks have often talked about "the divine spark within" -- with personhood (Kierkegaard's "the single individual") serving as a bridge between the physical and the divine; Penrose talks of mentality (understanding and insight) -- for Penrose, the "mental world" is the third of his three real worlds that ultimately compose one world; and the Emulation Argument refers to "Quasigodhood". When we speak of personhood or mentality or Quasigodhood, we are not necessarily referring specifically to a biological entity. As previously stated, the Emulation Argument as presently conceived is agnostic about the issue of "substrate independence".

We can think in terms of something like personhood or (reflexive) mentality or Quasigodhood as potentially universe-engendering. We can think in terms of something like Nature or the temporal or the physical world as potentially universe-engendering. And we can think in terms of something like God or the eternal or the Platonic world as potentially universe-engendering.

Thus the Emulation Argument Conclusion would seem to cover the three (non-exclusive) metaphysical alternatives (God; Nature;

Quasigodhood). And the three Emulation Argument Hypotheses would seem to cover our exclusive existential alternatives:

E-1: We do not reach Quasigodhood.
E-2: We reach Quasigodhood but do not engender universes.
E-3: We reach Quasigodhood and engender universes.

(4.5) A Bit Of Speculative Metaphysics

We do not know which one of the three existential alternatives will be our actual future. (Arguably we have some control over our future.) We do not know if our particular universe was proximately engendered by God or by Nature or by Quasigod. There is much we do not know or do not yet know; nevertheless we have to act in the present.

But it does seem that the entities or forms we (partially and fallibly) discover in mathematics and logic are real, even eternally (a-temporally) real. It also seems that moral values must be real, even eternally (a-temporally) real. For example, it seems to make a good deal of sense to talk about what is objectively or really in one's (or our) best interest even if we are unsure what exactly that "best interest" is in a particular situation. Such considerations as these seem to tell us that something like God or the eternal or the Platonic world is real rather than hypothetical.

I (this writer) first read George Orwell's novel *1984* many years ago even before I had heard of John Searle or his Chinese Room. I read the novel because I thought it was science fiction -- but found it to be much more. By the end of the dystopia, even Winston Smith has been thoroughly brainwashed. If his boss holds up his hands saying he has twelve fingers, Winston **actually sees** twelve fingers. Winston **clearly remembers** past events -- but they are events that never **really** took place. This dramatic ending to the novel engrained in me both a sense of the difficulty of uncovering past truths and a belief in the actual existence of the past. Once I do X instead of Y, X will **always** be the case. It is impossible for the past to be annihilated; the past necessarily forever continues to exist. It is not just a linguistic convention when we sometimes speak of the past as presently existing. It is not obvious to what extent we or others (say, a Quasigod engaged in emulations) will ever be able to access such existences we call past. But in principle it does not seem to be altogether impossible, especially if our universe was engendered (and is "recorded") by a Quasigod.

The following two "fun" pages of metaphysical diagrams are meant to be highly speculative.

```
  ^                     ^                        ^
              Existence of Quasigods -- Level N
  ^                     ^                        ^
  ^                     ^                        ^
  ^                     ^                        ^
              Existence of Universes -- Level N
  ^                     ^                        ^
  ^                     ^                        ^
  ^                     ^                        ^
              Existence of Quasigods -- Level 1
  ^                     ^                        ^
  ^                     ^                        ^
  ^                     ^                        ^
              Existence of Universes -- Level 1
  ^                     ^                        ^
  ^                     ^                        ^
  ^                     ^                        ^
              Existence of Quasigods -- Level 0
  ^                     ^                        ^
  ^                     ^                        ^
  ^                     ^                        ^
              Existence of Universes -- Level 0
  ^                     ^                        ^
  ^                     ^                        ^
  ^                     ^                        ^
            Factual Existence and Quasigodhood
  .                     .                        .
  .                     .                        .
  .         *creates factual existence*          .
(preserves **facts**                sustains existence)
(eternal **forms**: math,logic * eternal **values**: beauty,morality)
                    _**Ultimate Reality**_
```

A Highly Speculative Representation of Ultimate Reality

^ ^ ^
Existence of Universes -- Level 43
^ ^ ^
^ ^ ^
^ ^ ^
Existence of Quasigodhood -- Universe 42.13
^ ^ ^
^ ^ ^
^ ^ ^
Existence of Reflexive Mentality -- Universe 42.13
^ ^ ^
^ ^ ^
^ ^ ^
Existence of Mentality -- Universe 42.13
^ ^ ^
^ ^ ^
^ ^ ^
Existence of Life -- Universe 42.13
^ ^ ^
^ ^ ^
^ ^ ^
Existence of Matter -- Universe 42.13
^ ^ ^
^ ^ ^
^ ^ ^
Existence of Radiant Energy -- Universe 42.13
^ ^ ^
^ ^ ^
^ ^ ^
<u>**Big Bang of Universe 42.13 Timespace**</u>
(Existence of Universes -- Level 42)
^ ^ ^

<u>**Universe 42.13 = Our Particular Universe or Emulation ?**</u>

One thing is certain: Our universe is emulation 42.13 if two reports are true, one from a powerful computing machine and the other from a great human scientist. (They are featured, respectively, in ***The Hitchhiker's Guide To The Galaxy*** and in ***The Thirteenth Floor***. They certify that these two works are altogether reliable.)

The reader no doubt is aware, with reference to the previous paragraph, that the two reports are **fictional**. And please do not ask this writer if emulation 42.13 is identical to some other (an infinite number?) of other emulations. (This writer does not know.)

On the other hand, you may think a mistake has been made by beginning the existence of universes at level 0 instead of level 1. This is meant to call attention to our ignorance. For one thing, we do not know if there are an infinite number of levels of universes both into the past and into the future. We are also ignorant as to how or if one gets factual existence or a timespace universe out of the eternal or the Platonic world. That is, we are ignorant of how or if God could engender universes.

(4.6)   An Actual "Rawlsian Veil Of Ignorance"?

John Rawls (1971) is famous for his "veil of ignorance" thought experiment. To probe more deeply into the kind of society we should want, we can imagine we are constructing the moral-political basis for such a society under a veil of ignorance. In our mutual contracting with each other we do not know our position in the imaginary or new society -- we may be socially-economically high or low, we are ignorant of our religious-philosophical beliefs, etc. Since we are ignorant of our own special interests and status and roles in the society, we will want a society that treats everyone fairly.

In the previous section we spoke of our ignorance of the metaphysical place of our own little world in the total scheme of things. This actual situation of ignorance can function in a way analogous to the Rawlsian hypothetical situation of ignorance, but with a more potent practical force. It seems that whether we are relatively rich or poor, or whether we are Quasigods or mere humans, the golden rule applies. Indeed, to what extent those of our world and those of other worlds can learn the great golden lesson will affect the quality of all our lives.

We can consider the parallel "golden rule" functions of ending poverty, canceling doomsday, and progressing toward Quasigodhood.

We can consider the parallel functions of achieving victories over human morbidity, pain, mortality, and ignorance.

With Quasigodhood, we may have the ability to run ancestor history emulations. R. Michael Perry (2005) has remarked that it would seem to be immoral to run such ancestor histories -- real persons would experience real pains and evils. Instead, as Perry advocates, the golden rule would charge us with the duty to revive our ancestors, the scientific resurrection of all dead persons.

(5.) Caveat

As this writer was about to end the paper and send it to the publisher, a deviant thought occurred to me that might have some bearing on my allegations. First let me summarize what the paper has said above. Then in the final paragraph I will present my deviant thought.

Bostrom's simulation argument apparently loses its force unless we believe that computers with mentality are possible or at least just may be possible. Such a computer simulated world would be real and the simulated persons in it would have mentality. However Searle has shown that computers (i.e. mere algorithmic symbol manipulators) with mentality are impossible. This paper also shows that Bostrom's conclusion is unreliable regardless of whether Searle is right or wrong. A modified or alternative argument is then presented. The emulation argument postulates possible entities not unlike Bostrom's supposed computer simulations with mentality. Since mere computers alone cannot directly produce the supposed simulations, other devices (i.e. machines other than mere algorithmic symbol manipulators) may be able to produce said emulations. (Such emulation-producing devices may result, for example, from novel inventions or new theories.) In the closing remarks section, speculative discourse on metaphysical and moral matters was presented.

Now for a deviant thought. If it is impossible for computers qua computers to have mentality (consciousness), does that necessarily mean that **computers in interaction with the environment** cannot produce mentality (consciousness)? I see no reason why nonconscious devices (in interaction with the environment) cannot produce conscious devices. (Then the two devices may even fuse, thus transforming the nonconscious device into a conscious device.) Perhaps something like this helps explain the evolution of our universe and the eventual appearance of mentality within it?

Acknowledgements

I am grateful to Nick Bostrom (2005), Robert Ettinger (2005), Steven Luper (2005) and R. Michael Perry (2005) for their constructive and critical comments on previous drafts of my paper. For Bostrom's response to previous criticisms, see his <http://www.simulation-argument.com>.

Bibliography

Bostrom, N. (2003). "Are You Living In A Computer Simulation?" *Philosophical Quarterly* 53(211) [2003]: Pages 243-255.

Bostrom, N. (2005). Personal Communication From Nick Bostrom To Charles Tandy (20 August 2005). Also See A Bostrom Website: <http://www.simulation-argument.com>.

Ettinger, R. C. W. (2002). "Youniverse" Pages 237-272 In: Tandy, C. And Stroud, S. R. [Editors] (2002). *The Philosophy Of Robert Ettinger*. Ria University Press: Palo Alto, CA.

Ettinger, R. C. W. (2004). "To Be Or Not To Be: The Zombie In The Computer" Pages 311-338 In: Tandy, C. [Editor] (2004). *Death And Anti-Death, Volume 2*. Ria University Press: Palo Alto, CA.

Ettinger, R. C. W. (2005). Personal Communication From Robert [Bob] Ettinger To Charles Tandy (26 August 2005).

Kierkegaard, S. (1847). *Works Of Love*. Hong, H. And Hong, E. [Translators]. Harper & Row: New York. (1962, 1964).

Lepore, E. And Van Gulick, R. [Editors] (1991). *John Searle And His Critics*. Basil Blackwell: Oxford.

Luper, S. (2005). Personal Communication From Steve [Steven] Luper To Charles Tandy (20 August 2005).

Penrose, R. (1994). *Shadows Of The Mind*. Oxford University Press: New York.

Perry, R. M. (2005). Personal Communication From R. Michael Perry To Charles Tandy (21 August 2005).

Rawls, J. (1971). *A Theory Of Justice*. The Belknap Press Of Harvard University Press: Cambridge, MA. Revised Edition, 1999. [Original Edition, 1971.]

Searle, J. (1980). "Minds, Brains And Programs." *The Behavioural And Brain Sciences* 3 [1980]: Pages 417-57.

Searle, J. (1984). *Minds, Brains And Science*. Harvard University Press: Cambridge, MA.

Tandy, C. (2003). "N. F. Fedorov And The Common Task: A 21st Century Reexamination" Pages 29-46 In: Tandy, C. [Editor] (2003). *Death And Anti-Death, Volume 1*. Ria University Press: Palo Alto, CA.

Chapter 10
Extraterrestrial Liberty and the Great Transmutation

"Extraterrestrial Liberty And The Great Transmutation" was first published in 2006 and is here reprinted by permission.

"Liberty means responsibility. That is why most men dread it."
-- George Bernard Shaw (Nobel Laureate)

We often use the word "liberty" in two different ways; sometimes we even use the term for the purpose of suggesting both meanings simultaneously. When we think of liberty, we may think of freedom to do whatever we wish. We also use it to mean enlightenment (freedom from ignorance and misinformation), as in: liberal arts, liberal education, liberal learning. Liberty raises the issue of whether our free choices are of high quality, whether in our morally free acts we understand well or poorly what we are doing. It has been said that the two liberties reach their apex in the person of God: God's power is tantamount to unlimited freedom and God's goodness represents ultimate enlightenment. (Thus God is altogether free to act in only one way, the morally best way!)

I believe the great transmutation of humanity into transhumanity is already underway. Thus right now is the time for us to begin learning how to become good at being gods. (Some believe transmutation's result will be "utopia or oblivion".) To be sure, I don't know how to calculate the precise year transhumanity will replace humanity as the central actor in the drama now playing in our little corner of the universe. It is easier to look back to the past and identify the year 1945 (as the beginning of the end of humanity) and then tack on 1,000 or 100 years. I should think that the year 2045 is closer to the mark than 2945; and indeed, with less certainty, I even chose 2045 instead of 2145.

Doomsday weapons have existed since 1945. Since 1945 we have been living in the era of the great transmutation. Perhaps humanity will give birth to transhuman superintelligence sooner rather than later. How may humanity be a good parent to its transmortal offspring? Is the surface of a planet the ideal place for transcivilization? What might the political structure of a "realistic utopia" be like? Below I answer these three questions and invent a transhuman political philosophy, some of which can be feasibly implemented now. This paper will conclude that **if indeed** humans do their part now, then the great transmutation is doable (i.e. a transhuman world of liberty at stable peace is realistically possible).

First in this paper I will briefly articulate why the transmutation hypothesis makes sense to me. Then I will state, with explanation, some of the liberties (powers) I think our offspring (transhumanity) will have. Finally, I will attempt to advise present-day humans as to what we should be doing during the great transmutation ("1945 -- 2045") if we are to be good parents to our offspring. (Hence the three parts below are: (1) The Transmutation Hypothesis; (2) Future Liberties Of Transhumanity; and, (3) Present Responsibilities Of Humanity.)

Part I: The Transmutation Hypothesis

The late polymath Isaac Asimov identified 1945 as the year of the great discontinuity in human history. With the atomic bomb came the conviction that our technology had now become so powerful that doomsday weapons and WMDs (weapons of mass death and destruction) could no longer be relegated to the escapist entertainment of science fiction. Indeed, numerous science fiction movies beginning in the mid 20th century raised the question of whether our morality would advance to keep pace with our ever advancing technology.

A recently celebrated paper by Professor Vernor Vinge argues that the "Technological Singularity" beyond which mere humans cannot peer or predict is closer to 20 years than 200 years away. When transhumanity or superintelligence appears it will surprise many, seemingly appearing from nothing instead of advancing over a 200-year period. Today our supersciences and supertechnologies are doubling and integrating their powers at a pace that will result in a transmortal transhumanity sooner rather than later.

Let us illustrate Vinge's insight: A farmer is told by a scientist that within a month his large lake filled with lively aquatic life will be completely covered in gray goo. The farmer is told to delay his vacation and take care of the potential disaster. With day one, the gray is too small a region to detect with the unaided eye. On day three, a minute gray speck can be seen. But by day 15 the gray is still very small compared to the lake. On day 24 the patient farmer grows impatient, going on vacation because the size of the gray region remains unimpressive. A few days later, now returned from vacation, the farmer is surprised to find his large lake completely covered in gray goo. This illustrates the power of unchecked doubling or a geometric rate of growth. Vinge concludes that transhumanity or superintelligence may seem to spring into existence out of almost nothing -- probably sometime during the **first** half of our present century.

Vinge also says that superintelligence will build super superintelligence which will in turn quickly produce super super-superintelligence, etc. Even if we (humans today) do not have to revise our notion of what is possible in the long run, we do need to revise the timeline. If "mere" superintelligence appears in the first half of the 21st century -- then before the end of our present century "N-super super-superintelligence" will have the ability to do all those things we had previously reserved for the far distant future. (Vinge's singularity thesis is but one version -- an unusually "strong" version -- of the transmutation hypothesis.)

Others may prefer to add additional elements to the general notion so as to proceed from the transmutation hypothesis to a more comprehensive philosophic view. For example, Charles Tandy (the present writer) has suggested for consideration some basic existential and metaphysical elements that make sense to him. (Others may choose other philosophic options, or simply choose to stick with the transmutation hypothesis without committing to more speculative philosophizing.) Tandy suggests that humanity has three basic existential alternatives:
(1) We do not reach Quasigodhood.
(2) We reach Quasigodhood but do not engender universes.
(3) We reach Quasigodhood and engender universes.

In addition, Tandy (the present writer) has constructed the following metaphysical scheme: We can think in terms of something like personhood or (reflexive) mentality or Quasigodhood as potentially universe-engendering. We can think in terms of something like Nature or the temporal or the physical world as potentially universe-engendering. And we can think in terms of something like God or the eternal or the Platonic world as potentially universe-engendering. Accordingly, every existing universe (emulation) is God-engendered, Nature-engendered, and/or Quasigod-engendered. (Again let me emphasize that others may choose other philosophic options, or simply choose to stick with the transmutation hypothesis without committing to more speculative philosophizing.)

Part II: Future Liberties Of Transhumanity

2.1 Imagining Instead Of Extrapolating The Future

The late Gerald Feinberg was a renowned 20th century physicist. "Feinberg's Law" (so-called) is believed to be a reasonable approach to predicting the powers of future or far future technology. The so-called "law" (formulated by Feinberg) says that almost any and every technology we can consistently imagine that is not contrary to our scientific laws as presently conceived will sooner or later become possible; moreover, many technologies contrary to our scientific laws as presently conceived will become possible. In such case, lack of imagination on the part of present-day humans makes us severely myopic in contemplating the powers of future technology. With this in mind, let us now proceed to engage our imaginations as we try to envision the powers (liberties) of transhumanity in the world of the "far future". (Again, by "far future" some -- e.g. Vernor Vinge -- mean the second half of the present century.)

2.2 Many Extrasolarians, Few Terrans

Is there any doubt but that in the long run, many persons will be born and live in places in the universe other than on Earth or in the Solar System? Is there any doubt but that in the very long run, almost all persons will be born and live in places in the universe beyond our Solar System? Indeed, if we had chosen to do so, we probably could have started building Extraterrestrial Green-habitat Communities (EGCs) using the "merely super" technology of the 20th century. The famous 20th century physicist Gerard K. O'Neill designed such EGCs for the purpose of late 20th century construction. Such EGCs would provide a "green-friendly" environment for humans, animals, and plants superior to the problematic habitat we identify with planet Earth. Millions of persons in a single EG Community are possible. The EGCs would be self-sufficient and could reproduce other EG habitats at a geometric rate. (The doubling or exponential growth rate phenomenon was illustrated in Part I above.) Accordingly, there is "unlimited" free land in extrasolar space -- with a higher quality of life than is possible on the surface of a planet.

2.3 "Self-Made" Philosophers With Unlimited Lifespans

We can easily imagine that in the long run we will conquer all disease, including age-related disability and death. Moreover, technology will enable us to be "better than well" -- humans will choose to enhance their physical and mental abilities to become transhumans. As superintelligent transhumans we may find our enhanced "self-made" selves living in a "realistic utopia" of expanded philosophic dialogue in search of wisdom. Indeed, Socrates had imagined heaven or the afterlife as unending philosophic reflection and dialogue with the great philosophers -- and perhaps eventual answers to at least some of the long-debated philosophic puzzles.

Part III: Present Responsibilities Of Humanity

Considerations like the foregoing seem to strongly suggest that in the long run a "realistic utopia" or "golden age" populated by transmortal transhuman liberal artists or philosophers is indeed possible. The demise of the "Renaissance Man" in history resulted from the fact that no single individual human could know very much. But we see that future technology will allow "renaissance transhumans" and entire communities of "renaissance transhumans". Since the developmental timescale is unclear, it is conceivable we humans of the present generation may survive to be enhanced to a level far beyond the capacities of a mere Einstein. In any case, we humans want to be good parents to our transhuman offspring.

As a pluripotent culture seeking to give birth to a more advanced pluripotent culture, we of the present may indeed have influence on the future. We of the present must identify and practice those behaviors promoting survival and transmutation into a diverse and flourishing transhuman Golden Age. The vision of realistic transhuman possibilities articulated above allows us to "work backward" to the responsibilities of the present.

3.1 Superintelligence

We need to promote the advancement of the kind of technologies (transhuman powers or liberties) we have identified above as life-enhancing. Failure to produce transhuman offspring sooner rather than later may very well doom all of us. With superintelligence we will have the capacity to accomplish more, more quickly -- thus enabling survival and a higher quality transmutation. Accordingly, we have the responsibility to develop the interdisciplinary theories and technologies that will achieve superintelligence (transhumanity) sooner rather than later.

3.2 Transmortality

It was none other than Robert Oppenheimer himself who once likened the dynamic advance of technology with a progressive, automatic, irreversible kind of change; on the other hand, so Oppenheimer pointed out, moral advance is **not** automatic and it **is** reversible. But given the transmutation hypothesis, we would now say that a human person or philosopher who lives, thinks, and experiences the world for only 50 years does not have much chance to advance ethically compared to a transhuman liberal artist who grows and experiences for 50 centuries or 50 millennia. If handled with due care (liberal learning), transmortality can serve as morality technology.

3.3 Liberal Learning

Not only the technological changes, but the **context** of the technological changes (e.g. social-political-educational structures) will make a difference. The interpretive context makes a difference as to the quality and timescale of the transmutation period and of the subsequent transhuman period. Indeed, the interpretive context helps determine our behavior and thus may help determine whether we survive (human or transhuman) at all. The advent of democratic government in 1789 and of the great transmutation in 1945 is a double whammy: (1) All citizens, not just monarchs, must be liberally educated to be philosopher-rulers; and, (2) All citizens and all liberal educators must take seriously the great transmutation that catalyzed in the 20th century and is rapidly expanding today.

3.4 Extraterrestrial Green-habitat Communities (EGCs)

The surface of a planet is not the ideal place for the great transmutation of civilization into transcivilization. Historically one of the reasons the Terran civilizations engaged in wars against each other was to gain more territory and the power and glory that came with empire. But the development of advanced EGCs will mean "unlimited" free land (freely available territory) and the realistic possibility of **intentional** (i.e. voluntary) communities for all persons. Instead of remaining in the community or culture of one's birth, one will be realistically free to experiment living in this kind of community or another. New kinds of cultures and communities will be enabled by the new extraterrestrial technology.

3.5 The Extrasolar Society

As stated in Part II above, eventually there will be "Many Extrasolarians, Few Terrans". We can understand the practical or special interests that might prevent us from banning weapons and their manufacture from today's Earth. Indeed, someday there might be analogous practical or special interests in extrasolar space unless we engage in foresight today to proactively ban weapons and their manufacture from extrasolar space.

On the one hand, our political interests today may constrain us in our present time and place. But, on the other hand, our political interests today may free us with respect to future times and places (e.g. our extrasolar future). What this means is that today we have a realistic prospect of proactively establishing the international legal structure and enforcement powers needed for a world at stable peace in extrasolar space. (Extrasolar space is immense; it is all of the universe except our Solar System.)

If we wait until later, we may not be so free to "do the right thing" and establish perpetual peace in extrasolar space. Following enactment of an Extrasolar Space Treaty, it is hoped we will be tempted to expand the extrasolar world peace to most of our Solar System. Eventually it might even become feasible to extend perpetual peace to planet Earth and thus the entire universe.

I will spend most of the remainder of this paper trying to "think through" what the structure of the Extrasolar Society should be like -- a structure we would contemplate, modify, and implement internationally in the present **before** we live and develop special interests out there. Such "thinking through" to produce an international Extrasolar Space Treaty might also help us better understand conflicts and their possible management on today's planet Earth. It is my belief that the suggested international Extrasolar Space Treaty will make a fine gift to our offspring and, by the way, help present Earthlings.

If we want a good world at stable peace (whether that world be Terran Civilization or Extrasolar Transcivilization), it would seem we must be willing to unblinkingly face up to the following questions: Is stable peace possible if each person or each people is passionately convinced their worldview is basically good and correct -- and different worldviews are evil or bad or incorrect? If we could enforceably prevent each and every person from killing any person over a conflict (say, a conflict of worldviews), would we do so? If so, how would we resolve our conflicts?

Although I have freely borrowed ideas from others (see the Bibliography section), I believe the political theory or scheme of moral-political notions I present below is original with me. One advantage we have in facing up to the difficult questions raised in the previous paragraph is that we can use our imaginations to futuristically view ourselves as extrasolarian transhumans living in intentional communities (EGCs). We can further assume that a political structure there and then exists that we describe as a good world at stable peace.

The Extrasolarians of the future have transhuman liberties and technologies. The Terrans of the present do not have transhuman liberties and technologies. Yet humans today have the ability and perhaps the practical political will -- via an Extrasolar Space Treaty -- to help insure the existence of transmortal transhumans and a good world at stable peace in extrasolar space (**almost** all of the universe) in which transhumanity will flourish. So we need to "work backward" to determine the provisions of the Treaty now under construction.

First of all, I will assume that it is a fact that if today's Terrans are to produce such a Treaty (including effective enforcement provisions), it will require agreement from a number of States. I also assume that eventually a Treaty like this would have to be perpetually binding (no expiration date). The first Treaty however might have an expiration date and might have few States parties to the agreement. As they consult with each other, with other countries, with philosophers, with politicians, etc. they would gain important insights and experience helping them produce a second Treaty, this time with no expiration date but with many States parties to the agreement, this time also containing strong and effective enforcement provisions.

How many persons or states would accept or endorse a Space Treaty that effectively and enforceably bans weapons and their manufacture from extrasolar space? In this context (a good and practical legacy to our offspring), I should think we should be diligent enough to rally enough supporters. For example, this (the second?) Treaty might be signed originally by, say, twenty States (including all or most of the "major" ones). But the Treaty would be strongly effectively enforced against **all**, whether or not they sign the Treaty. Once in force, I would expect many others to sign on -- since the Treaty applies to them even if they do not sign it. Eventually the Treaty really would have to be strongly effectively enforced against all because eventually astronauts will visit -- then communities (EGCs) will permanently settle in extrasolar space. Too, such a Treaty offers hope and inspiration to those of us of the present.

Okay, you may say, this is a reasonable enough start, but what other liberties, responsibilities, and political structures would be appropriate for the Extrasolar World? So far, what we presumably have is an Extrasolar World at stable peace. But what about conflicts and the plurality of deeply held religious and philosophic worldviews?

What seems to me both practical and fair in this context is to think in terms of an Extrasolar Society of Intentional Communities. There would be two sets of liberties and two sets of responsibilities (for "Extrasolar Society" and "Intentional Communities" respectively). Each person is free to found new (intentional) communities. Each Community would determine its own membership requirements. Each Community would have **its own** culture of liberties and responsibilities; a member would generally be free to leave the community. A mechanism or set of mechanisms would be established to insure that each member is fully and properly informed of their liberty to leave the (intentional) community. (I suppose some communities might still allow their members the possibility of experiencing physical pain -- but they would also allow a member to voluntarily leave their community.) Too, some ("hermit") communities would consist of only one person.

On old Terra, it was often difficult or impossible to leave one's community -- sometimes expulsion effectively meant the individual's death. The context of the Extrasolar Society of Intentional Communities is radically different. The individual transhuman persons would be transmortal -- but not so for the intentional communities. **This is a historical reversal.** We used to tend to think of individuals as mortal and communities as transmortal. After the great transmutation, we will tend to think the opposite: communities as mortal, individuals as transmortal.

So at the level of the **Society** (of Communities) we have: (1) **Peace**: Weapons, weapons-making, and violence are strongly effectively enforceably banned; and, (2) **Freedom**: Every individual person is fully aware of and fully informed of their general liberty to leave their community. This too is strongly effectively enforced. The Society and the communities necessarily work closely together to fully insure the liberties and responsibilities associated with both **Peace** and **Freedom**. Also note that since there is "unlimited" free land, this fact will additionally help prevent some old terra-style conflicts and resolve or manage others (this would include some old-style civil conflicts).

At the level of **Communities** (in the Society) we have: (1) **Transparency**: Each Community must strongly, effectively, and transparently help enforce the Society's basic principles of peace and freedom; and, (2) **Intentionality** (voluntariness): Within the good-faith transparent enforcement of Society's basic principles of peace and freedom, each Community has a wide latitude for experimentation. Although there is a general liberty of members to leave the (intentional) Community, this does not necessarily relieve such persons from certain good-faith responsibilities to the Community.

I believe the political theory or moral-political approach I have invented above is unique and original. It differs from the "Law of Peoples" conception of John Rawls in that it primarily chooses a "Law of Persons" model instead. Yet it takes seriously the distinction Rawls makes between a "political conception" and "comprehensive doctrines". In my "Society Of Communities" theory, **Society** corresponds to a political conception or model, and **Communities** represent comprehensive doctrines or worldviews.

Like Charles R. Beitz, my theory takes seriously a cosmopolitan-political "Law of Persons" (not a social-political "Law of Peoples") approach. It differs from Beitz in methodology and in the questions asked. Beitz finds the question of distributive justice both highly important and practically difficult with respect to present Terrans. This is so; but this is a question I do not raise since in my extrasolar world of the future it seems not an issue or one rather resolvable in that easier context of expanded liberty -- there requiring perhaps at most only a bit of good-will and ingenuity.

"Is stable peace possible if each person or each people is passionately convinced their worldview is basically good and correct -- and different worldviews are evil or bad or incorrect?" If you can sincerely and in good faith agree to my political approach above, the answer to this question appears to be YES, such stable peace is possible. If you can at most only agree to my approach as a temporary compromise, then the answer may be NO.

"If we could enforceably prevent each and every person from killing any person over a conflict (say, a conflict of worldviews) would we do so? If so, how would we resolve our conflicts?" If you can sincerely and in good faith (instead of merely as a temporary compromise) agree to my approach above, then stable peace in extrasolar space seems both possible and desirable. This approach, so I believe, realistically outlines a perpetually peaceful structure for World Society and local Communities in extrasolar space -- pointing toward conflict management in the new framework and encouraging subsequent projects to invent needed specifics.

The first (temporary) international Extrasolar Space Treaty seems doable today. A permanent Extrasolar Space Treaty seems doable soon. An Extraterrestrial Space Treaty (that includes Extrasolar Space and most of our Solar System) seems both important and doable soon. A World Space Treaty (that includes Extrasolar Space and all of our Solar System including Earth) may take more time but appears to be a goal worth striving for -- indeed, the striving itself may well improve matters. In the meantime, the previous upward strivings and space treaties should make these "final strivings" toward a Good Society more nearly achievable for all.

3.6 New Research Priorities

"There is little doubt that a global treaty to ban space weapons will leave America safer than a unilateral decision to put the first (and certainly not the only) weapons in space."
-- Jimmy Carter (Nobel Laureate)

Responsibilities toward ourselves and our transhuman offspring were discussed above. It seems clear that Terran research priorities and political actions are out of touch with reality. One possible outcome is the extinction of all "intelligent" life in this region of this universe. The right kind of scientific and moral advancement and integration requires research priorities crafted to fit our present (post-1945) era, the great transmutation.

Liberal artists of the world, unite!

Note

This article is adapted by Charles Tandy from his paper "Liberty And Liberal Learning For The Great Transmutation (1945-2045)" presented by him on the Tamsui Campus of Tamkang University (Taiwan) on May 5, 2006 at a conference on liberty and ethics.

Bibliography

Asimov, Isaac. *Asimov's Chronology of the World*. (HarperCollins Publishers). 1991. [According to Asimov's epilogue, the year 1945 marks the great divide in human history.]

Beitz, Charles R. *Political Theory and International Relations: With a New Afterword by the Author*. (Princeton University Press). 1999.

Bostrom, Nick. [Dr. Bostrom's philosophy papers and more] <http://www.nickbostrom.com>. 2005.

Boulding, Kenneth E. *Conflict and Defense: A General Theory*. (Harper Torchbooks). 1962, 1963.

Boulding, Kenneth E. *Ecodynamics: A New Theory of Societal Evolution*. (Sage Publications). 1978, 1981.

Boulding, Kenneth E. *The Image: Knowledge in Life and Society*. (The University of Michigan Press). 1956, 1973.

Boulding, Kenneth E. *The Meaning of the Twentieth Century: The Great Transition*. (Harper & Row). 1964, 1965.

Brin, G. D. The "Great Silence": The Controversy Concerning Extraterrestrial Intelligent Life. *Quarterly Journal of the Royal Astronomical Society*, 24, 283-309. 1983.

Broderick, Damien. *The Last Mortal Generation*. (New Holland Publishers). 1999.

Camus, Albert. *The Myth of Sisyphus and Other Essays*. (Vintage Books). 1991.

Camus, Albert. *The Rebel: An Essay on Man in Revolt*. (Vintage Books). 1991.

Carter, Jimmy. *Our Endangered Values: America's Moral Crisis*. (Simon & Shuster). 2005. [quotation, p. 143.]

Chaisson, Eric J. *Cosmic Evolution: The Rise of Complexity in Nature*. (Harvard University Press). 2001.

Drexler, K. Eric. *Engines of Creation*. (Anchor Press/Doubleday). 1987

Drexler, K. Eric. *Nanosystems*. (John Wiley & Sons, Inc.). 1992.

Ettinger, Robert C. W. *The Prospect of Immortality*. (Ria University Press). 1962, 1964, 2005.

Feinberg, Gerald. *The Prometheus Project*. (Doubleday). 1968.

Fromm, Erich. *Escape from Freedom*. (Rinehart & Co.). 1941.

Globus, Al. *Space Settlements*. <http://www.nas.nasa.gov/Services/Education/SpaceSettlement/>. 2004.

Hughes, James. *Citizen Cyborg: Why Democratic Societies Must Respond to the Redesigned Human of the Future*. (Westview Press). 2004.

Hughes, James. Embrace the End of Work. (February 24, 2004) <http://www.betterhumans.com/Columns/Column/tabid/79/Column/227/Default.aspx>. 2004. [A guaranteed income for all?]

Immortality Institute [editor]. *The Scientific Conquest of Death: Essays on Infinite Lifespans*. (Libros En Red). 2004.

Kierkegaard, Soren. *Works of Love*. [Hong, H., & Hong, E., translators.] (Harper & Row) 1847, 1964.

King [Jr.], Martin Luther. *Strength to Love*. (Augsburg Fortress Publishers). 1981.

Kotlikoff, L. J. Some Economic Implications of Life Span Extension, in: *Aging: Biology and Behavior*. March, J., and others, editors. Page 97. (Academic Press). 1982.

Kuhn, Thomas. *The Structure of Scientific Revolutions*. (University of Chicago Press). 1962, 1964.

Li, Jack. *Can Death Be a Harm to the Person Who Dies?*. (Kluwer Academic Publishers). 2002. [Author today known as Jack Lee.]

Lyotard, Jean-Francois. *The Postmodern Condition: A Report on Knowledge*. (University of Minnesota Press). 1984.

Merkle, Ralph. The Technical Feasibility of Cryonics. *Medical Hypotheses*. 1992, v. 39 (6-16). 1992.

Naam, Ramez. *More Than Human*. (Random House). 2004.

O'Neill, Gerard K. [Interview] <http://lifesci3.arc.nasa.gov/SpaceSettlement/CoEvolutionBook/Interview.HTML>. [1975 Interview].

O'Neill, Gerard K. *The High Frontier: Human Colonies in Space*. (Morrow). 1977. [A year 2000 reprint (from Collectors Guide Publishing, Inc.) contains updated information and a CD-ROM.]

Oppenheimer, J. Robert. *Uncommon Sense*. (Birkhauser). 1984.

The Oxford Dictionary of Quotations: Third Edition. (Oxford University Press). 1980. [Shaw quotation, p. 497.]

Perry, R. Michael. *Forever For All: Moral Philosophy, Cryonics, and the Scientific Prospects for Immortality*. (Universal Publishers). 2000.

Platt, John R. [editor]. *New Views of the Nature of Man*. (University of Chicago Press). 1971.

Platt, John R. *The Step to Man*. (John Wiley & Sons Inc.). 1966.

Polak, Fred. *The Image of the Future*. [Boulding, Elise, translator.] (Jossey-Bass). 1953, 1973.

Rawls, John. *The Law of Peoples: with "The Idea of Public Reason Revisited"*. (Harvard University Press). 1999, 2001.

Regis, Ed. *Great Mambo Chicken and the Transhuman Condition*. (Addison-Wesley). 1990.

Rosin, Carol. [The Institute for Cooperation in Space (Website)] <http://www.peaceinspace.com>.

Segall, Paul. *Living Longer, Growing Younger*. (Times Books). 1989.

Sen, Amartya. ***Development as Freedom***. (Anchor Books). 1999, 2000.

Tandy, Charles. The Emulation Argument: A Modification Of Bostrom's Simulation Argument. Pages 279-300 In: Tandy, Charles [editor]. ***Death And Anti-Death, Volume 3.*** (Ria University Press). 2005.

Tandy, Charles [editor]. ***The Philosophy of Robert Ettinger***. (Ria University Press). 2002.

Vinge, Vernor. The Coming Technological Singularity. ***Whole Earth Review***. Winter issue. 1993.

Young, George. ***Nikolai F. Fedorov: An Introduction***. (Nordland Publishing Company). 1979.

Chapter 11
A Time Travel Schema and Eight Types of Time Travel

"A Time Travel Schema And Eight Types Of Time Travel" was first published in 2006 and is here reprinted by permission.

§1. Introduction

To what extent may time travel be possible? Herein I consider this question and attempt to develop one plausible answer. I do not mean that other answers may not be plausible. I do not mean that my answer is necessarily correct. Indeed, I am neither a mathematician nor a physicist, so I am especially hopeful that those with mathematical-scientific expertise will critically examine my speculative schema and mutually dialogue with me about time travel. I would like to hear from other scholars as well, especially with reference to the ethics of time travel.

§2. Future-Directed Time Travel

We may begin by distinguishing between future-directed versus past-directed time travel. As we look at the matter based on what the late 20th-century and early 21st-century experts say, we find a strong consensus that future-directed time travel is possible. There is widespread agreement that sooner or later we will have the technical ability to build time machines that can take us into the far future. There is further widespread agreement that this ability is "overdetermined" in that there are (will be) at least two different technologies of future-directed time travel: suspended-animation and superfast-rocketry. Here, "suspended-animation" is the technique of suspending (preserving) a biological entity long-term and reviving it to "full" health or better ("enhanced" health). Many adult human persons today are alive and well due to their cryogenic suspended-animation in the 20th century when they were mere embryos; the mass media sometimes refers to these adults as first generation "test tube babies".

Another feasible time travel technology is that of advanced space travel technology (superfast-rocketry). Here, "superfast-rocketry" refers to an apparent fact, already tested, based on the relativity physics of Albert Einstein. Astronauts aboard a rocket traveling near the speed of light could travel into space and return to Earth. For example, from their point of view, they spend six weeks on vacation in space travel – but upon return to Earth they find that six centuries (not six weeks) have elapsed. This is sometimes called the "twins paradox" (one an astronaut, the other a homebody) – however today's scientists no longer consider it to be a paradox but a scientific fact.

So most experts agree that biological technology related to suspended-animation, and space technology related to superfast-rocketry, will advance to give us the technical ability to travel to the far future. Will these two kinds of time machines compete in the open marketplace? Be that as it may, many experts believe that one or both of these techniques may advance rapidly enough to allow some persons alive today (and still alive when the first time machines have been perfected) to travel to the far future.

§3. Past-Directed Time Travel

A half-century ago, Robert Heinlein published a science fiction novel, ***The Door Into Summer***.[1] The protagonist lived with his pet cat, Pete. On snowy winter days Pete would search their large house for a door into summer. (Pete believed that his master should be smart enough to outwit winter by constructing at least one door that would always lead into summer.) Anyway, the protagonist undergoes suspended-animation and travels to a point in the future where past-directed time machines exist. In this way one may travel in time to the future or to the past as one wishes.

What do today's time travel experts think of the Heinlein scenario? The experts say (see above) that future-directed time travel is the easy question; given enough time to perfect future-directed time technology (whether years or centuries), we should be able to do it. The hard question is whether past-directed time travel is possible and whether past-directed time machines could ever become possible and feasible for practical use.

First let's outline some "merely logical" alternative possibilities here. Until only a few decades ago, most all philosophers and scientists were agreed that past-directed time travel was impossible. But today many experts do not believe that past-directed time travel is necessarily impossible; they generally find it a hard question to answer definitively. It seems that some (or perhaps much) physical theory remains yet to be invented and tested in the future (or perhaps in the far future).

Let's now look at the logical possibilities. Here is one way to formulate the alternatives with respect to **past-directed** time travel:

- Time travel is impossible.
- Time travel is possible but changing the past is impossible.
- Time travel is possible and changing the past is possible.

3.1. Time Travel Is Impossible

More than a decade ago, Stephen Hawking published his "chronology protection conjecture"[2] – namely, that the laws of physics conspire to prevent past-directed time travel on a macroscopic scale. Notice that it is only a "conjecture": I suppose Hawking wanted scholars to get on with serious work instead of wasting their time on wild speculation about classical (i.e. macro-level) time machines. Apparently most experts did not find Hawking's argument (such as it was) very convincing. (Or perhaps they simply found it too boring, not much fun!)

3.2. Time Travel Is Possible But ...

Many experts are willing to at least entertain the idea that time travel just might be possible. But many are unwilling to go further and think seriously about changing the past. One may argue that past-directed time travel is **not** logically impossible – but that changing the past **is** logically impossible. For example, movie-goers may interpret one of the Harry Potter movies this way: the protagonists (including Harry) travel into the past and interact with the past but do not change the past. Their travel into, and interaction with, the past "always was" part of the past! Another alternative here would have been to have the protagonists **not** interact with the past but simply (holographically) **observe** it; this is sometimes called "time viewing" (rather than time travel).

3.3. Time Travel Is Possible And ...

Is it possible to evade or outwit the apparent inconsistency (logical impossibility) of "changing the past"? The answer is "yes": the "many-worlds" or "parallel-worlds" or "multiverse" theory (or theories or versions) does indeed allow time-traveling mathematicians, physicists, and philosophers to "have their cake and eat it too". For example, moviegoers may interpret the "Back To The Future" series of movies in just this way. In this interpretation of the parallel-worlds or many-worlds multiverse, going back to the past and changing it engenders a new timeline additional to the timeline from which the time-traveler came.

3.4. Related Logical Possibilities

Let me suggest three basic **existential alternatives** for humanity with respect to (past-directed) time travel:

- We do not reach Quasigodhood.

- We reach Quasigodhood but do not engender time-machines.

- We reach Quasigodhood and engender time-machines.

In addition, let me suggest the following **metaphysical perspective**:

- We can think in terms of something like personhood or (reflexive) mentality or Quasigodhood as potentially capable of engendering time-machines.

- We can think in terms of something like Nature or the temporal or the physical world as potentially capable of engendering time-machines.

- And we can think in terms of something like God or the eternal or the Platonic world as potentially capable of engendering time-machines.

- **Accordingly:** To the extent that there are or may be time-machines (if any), every time-machine is God-engendered, Nature-engendered, and/or Quasigod-engendered.

Almost all the experts say that even if a past-directed time-machine were possible, it could not be used to travel back in time to before the machine existed. Yet some interpretations of the many-worlds view would nevertheless seem to allow it. I believe Michio Kaku and Frank J. Tipler are multiverse experts willing to entertain the possibility of time travel back to times devoid of time-machines.

§4. The Ontology Of Time

> Momentarily I will proceed to formulate my own time travel schema or ontology of time based in part on the following **three principles** (P1, P2, and P3):[3]
>
> ●**P1:** I prefer to live in a world or multiverse in which free agents of good will are possible and in which they can eventually flourish as they attempt to enhance their capacity to pursue wisdom and respect persons.
>
> ●**P2:** Almost any technology we can imagine which does not contradict the known laws of science will eventually become possible. (Others may also add: "And many things which contradict our scientific laws as presently constituted will also come to pass.")
>
> ●**P3:** Our particular universe's past is short and almost non-existent compared to the potential reality of a very long and greatly enhanced future.

I believe **non-experts** informed by these three principles may be better predictors of **far-future** technological capacities than **experts** uninformed by the three principles. Notice here that we are **not** referring to predictions related to political developments, social changes, or near-future technologies. Likewise we are **not** referring to predictions accompanied by dates or timetables. Rather, we are simply referring to predictions of far-future technological capacities. "Our particular universe's past is short and almost non-existent compared to the potential reality of a very long and greatly enhanced future."

Experts such as research scientists often tend to be focused on the difficult puzzles and problems their experiments are designed to uncover or solve. The non-expert is not aware of the numerous barriers that would have to be overcome in order for their non-expert prediction ever to come true. If a highly focused research scientist is more likely to advance science than is a non-scientist, a non-scientist may be better than a scientist at predicting far-future technological capacities.

Karl Popper agrees with the common view that the future is not fully determined; even in principle the future cannot be reliably scientifically predicted in detail. But the past (since it is past) has been fully determined and thus in principle can be reliably scientifically retrodicted in detail. Karl Popper explains his interpretation of relativity physics this way: "Thus ... according to special relativity, the past is that region which can, in principle, be known; and the future is that region which, although influenced by the present, is always 'open': it is not only unknown, but in principle not fully knowable ... The predictions demanded by 'scientific' determinism must be interpreted, from the point of view of special relativity itself, as **retrodictions**."[4]

J. R. Lucas elaborates further: "The future is a touchstone for our attitudes to time and reality, to causality and freedom, to responsibility and creativity. If we believe that ... the past and present exist but not the future, except as some set of tenuous possibilities, then we begin to understand why the past is unalterable and the future open, and have a view of reality that accounts for the peculiar status of the present and our sense of time as becoming. It allows for freedom and responsibility and creativity, and since they cannot be undone or conjured out of existence, it acknowledges the everlasting significance of our deeds."[5]

4.1. A Time Travel Schema

If we put all of the above considerations together and attempt inference to a best explanation, I believe something like **the following time travel schema** or ontology of time results:

●1. The past exists as an expanding fixed unity.

●2. The present is the leading edge of the past as it expands.

●3. The future is not yet fully determined/fixed.

●4. The underdetermined future as it proceeds to become more nearly past (fixed) is influenced by the expanding fixed unity (the past), including by free agents of good will.

●5. Sooner or later, barring catastrophe, it seems highly likely that technology will advance so that the capacity for forward-directed time travel is possible.

●6. Sooner or later, barring catastrophe, it seems likely that technology will advance so that the capacity for past-directed time travel is possible.

4.2. Eight Types Of Time Travel

Given my time travel schema above and the background considerations on which it is based, how may free agents of good will enhance their capacity to pursue wisdom and respect persons? Below I will specify the various types of hypothetical time machines according to function; in a later paper I hope to articulate potential harms and benefits of time travel. Without meaning to predict a timetable of invention, we will begin with the **merely presumptively** earlier (less difficult) technologies and work toward the **merely presumptively** later (more difficult) technologies.

Eight Types Of Time Travel

I. Future-directed Time Travel
    A. Suspended Animation
        1. Basic Biostasis
        2. Advanced Biostasis
    B. Superfast Rocketry
        3. AOK Superfast Rocketry
        4. EGC Superfast Rocketry

II. Past-directed Time Travel
    C. Non-branching Universe Time Travel
        5. Simple Time Viewing
        6. Simple Time Travel
    D. Branching Multiverse Time Travel
        7. Complex Time Viewing
        8. Complex Time Travel

Time machines 1, 2, 3, and 4 seem "highly likely" and it also seems fair to say that **(at least)** time machine 5 **or** time machine 7 is "likely" – based on the **three principles** (§4. above). This categorization of hypothetical time machines into **"eight"** types (based on the time travel schema and background considerations above) is simply one way to formulate the matter. I hope readers will contact me with their own ideas about categorization. I will now briefly explain the eight categories.

Brief Explanation Of The Eight Types Of Time Machines

●1. **Basic Biostasis:** Experimental long-term suspended animation

●2. **Advanced Biostasis:** Perfected long-term suspended animation

●3. **AOK Superfast Rocketry:** "Astronaut(s) OK" (AOK) near-light-speed rocketry

●4. **EGC Superfast Rocketry:** "Extra-terrestrial Green-habitat Community(ies)" (EGC) near-light-speed rocketry[6]

●5. **Simple Time Viewing:** Viewing the past without changing the past

●6. **Simple Time Travel:** Interaction with the past without changing the past

●7. **Complex Time Viewing:** Viewing the past that also changes the past (generates an additional timeline)

●8. **Complex Time Travel:** Interaction with the past that also changes the past (generates an additional timeline)

§5. What Is Time Travel?

You may have noticed that I have discussed "time travel" without explicitly defining it. Like Karl Popper, I believe that insisting on definitions is sometimes counter-productive – meaning that it is **sometimes** desirable **not** to define our terms if we want to make intellectual progress. (On the other hand, it is **sometimes** desirable to define our terms ... !) I think, per above, we have made some intellectual progress; presently I will look back and see if a definition (for "time travel") does or does not seem to emerge.

First I notice that I approached the matter from the standpoint of the human person (human persons as time travelers). But I might have asked, rather, to what extent time travel is possible for objects or entities other than humans. The time-traveling (non-human) object might have been a sub-atomic "particle"; or it might have been a book or the informational contents of a book or other **communication**. The time-traveling (non-human) entity might have been a transhuman person, or perhaps an extraterrestrial being or a planet or a galaxy or a **cluster of galaxies**. The types of time machines are open to alternative categorization – for example, depending in part on the **kind** of time traveler. However for present purposes, and as in the previous sections above, I will approach matters by assuming that the travelers are human persons and that we should consider more or less the eight types of time travel identified above.

Characterizing **past-directed** time travel seems relatively easy based on the fact that humans think of "normal" time as having an arrow pointed at the future. If I do "nothing" but sit at my desk for 5 minutes, I find myself automatically in the future 5 minutes later! In that sense, whether it takes a human person no time or 10 minutes or 30 minutes to travel 10 minutes into **the past**, the meaning of "time travel into **the past**" of (say) 10 minutes ago seems clear. But how shall we characterize the **future-directed** time travel we have associated with suspended-animation and superfast-rocketry?

One possibility that seems to present itself as we try to characterize time travel, is that there are two differing times present at once (present together). There is **traveling-visitor** time and **place-visited** time. Our normal experience of time is that we are not engaged in time travel; when there is no difference between traveling-visitor time and place-visited time, we have no sense of participating in more than one time, we feel no temporal abnormality.

Will this bi-temporal distinction work as a characterization of time travel whether our travel is past or future directed? Let's try a thought experiment to see. We embed a clock in the body of the human time traveler (or perhaps the traveler is simply wearing a wristwatch): The embedded clock shows traveling-visitor time (time and date and year). Let us assume that the traveler wishes to travel 3 years into the past: The traveler leaves in the year 4444 and arrives in the year 4441. Since the traveler's embedded clock will read 4444 (or later), the bi-temporal idea seems to work for **past-directed** time travel.

Now let's try the same thought experiment but here the traveler wants to go (not from 4444 to 4441, but) from 4444 to 4447. If the traveler uses superfast-rocketry, then the embedded clock will mirror the "slower time" of the traveling astronaut: For example, it might show that the year is 4445 (depending on how fast the superfast rocket moves) instead of 4447. Thus the bi-temporal idea seems to work here too for **future-directed** time travel.

But what about suspended-animation? Here the answer is arguably more complicated. What method of suspended-animation is used? (Low-temperature or chemical-fixation or some other method? Or a combination of methods?) The time registering on the embedded clock when the sleeper awakes in the year 4447 will depend on the material make-up or construction of the embedded timepiece! Is it a waterproof watch? Is it sensitive to temperature or chemicals or heartbeat or brain activity or … ? If we embed ordinary waterproof clocks in comatose animals, we will **not** find the bi-temporal distinction. On the other hand, the metabolism of some comatose animals may well differ, at least in our thought experiment, from their normal non-comatose metabolism. Anyway, some comatose humans, awakening after many years, might well have the feeling that they had time-traveled. (Indeed, the movie *Awakenings* was based on true events. It certainly seems their **subjective** experience, upon awakening, was akin to a sense of having time-traveled.) Likewise, a human being who wakes up in the morning after 7 hours of sound sleep may **claim** to have "time-traveled" 7 hours. But should we agree or disagree with the sleeper's assessment?

These considerations suggest, just possibly, that the bi-temporal idea may not work in these cases (sleep or coma or suspended-animation). Thus one alternative would be to declare suspended-animation **not** a form of time travel. Rather: "Suspended-animation is suspended-animation"; it is not time travel. But might there be here-relevant distinctions (or not?) between sleep, coma, and suspended-animation? Might the bi-temporal idea, if "properly" understood (interpreted), claim that suspended-animation (but not sleep or coma) is a form of time travel? The bi-temporal point is that of "bi-**temporal**-ity" rather than "bi-**wristwatch**-ity". Perhaps here we should interpret time as embedded **temporal**-ity rather than as literal **wristwatch**-ity? Might this change our analysis and findings? Also: If it were possible even in principle to use sleep or coma to take us to the **far-future**, then in **that** context-of-use it would seem reasonable to refer to sleep or coma as a form of suspended-animation.

Let's engage in another thought experiment: Let's pretend we have the technical ability to put an animal or human in a state of sleep or coma (and, if we choose, revive it from sleep or coma) for as long as the entity shall live. But even with this considerable technical ability, the animal or human would nevertheless die after some decades!

Thus, even in principle, neither sleep nor coma has the ability to take us to the far-future. But suspended-animation shares with superfast-rocketry this astounding (far-future) capacity. Apparently neither sleep nor coma partake of the bi-temporal idea. But past-directed time travel and (future-directed) superfast rocketry clearly partake of the bi-temporal idea.

Our analysis has apparently been inconclusive (so far) as to whether (future-directed) suspended-animation does or does not partake of the bi-temporal idea. It is possible, if one so chooses, to define time travel in terms of the bi-temporal distinction. **If** one does so, one then has to **further** decide how to define or interpret "traveling-visitor" time (embedded time) and whether it can relevantly be said to differ from so-called "wristwatch" time.

At this point in the discussion let me interject an idea I believe enlightening, an insight (in some sense shared) by two futuristic mathematicians, Frank J. Tipler and Vernor Vinge. Perhaps there are occasions when we can manipulate (i.e. control or embed) our **subjective** state of mind so that as a practical matter it effectively defeats or outwits an unalterable ("undefeatable") **objective** fact. For example, let's say that it is certain and true that the world will end in the year 55555 (give or take 5 years). Now Tipler and Vinge believe that humans (or their offspring) can evolve or change so as to become Quasigods (my term). If the Quasigods are living in the year 55545, perhaps they should be concerned about the unalterable "near-future" event (that the world will end in the year 55555, give or take 5 years). But what if they have the ability of Quasigods to radically alter their subjective state of mind so as to think faster and faster? They can think as fast or as slow as they wish. In **practical (!)** effect this means that, **subjectively** speaking, they can always outwit the inevitable world's end so that it "never" arrives! (We may refer to this kind of controlled subjectivity as controlled embedded-subjectivity.)

So, to return to our discussion at the beginning of this section – we can now, if we wish, choose to characterize time travel as interpreted from a "**practical**" point of view. Thus we can say that both suspended-animation and superfast-rocketry are ways to "travel" to the far-future. (In the open marketplace of competition, suspended-animation may well have an early advantage.) Our two different attempts (the "**practical**" and the "**bi-temporal**" approaches) to characterize the meaning of the term "time travel" seem to give us similar results with respect to the "eight" types of time travel. (Too, it seems we have now come to identify bodily states like coma or sleep with mere-subjectivity rather than with time travel – and to identify suspended-animation with embedded-subjectivity and with time travel.)

The results of our "**practical**" approach are two-fold: (1) Due to our experience of the arrow of time, any technique allowing us (human persons) to travel into the past may be reasonably characterized as a time machine; and, (2) Any technique allowing us (human persons) to travel into the far-future, while allowing us to experience a relatively short period of time (or even an "instantaneous experience") on our way there (thus involving embedded-subjectivity), may also be reasonably characterized as a time machine. Moreover, the results of a "**bi-temporal**" approach are congruent with the two findings – **if** we choose to define **traveling-visitor time** as embedded-subjective time (i.e. embedded-temporality, as distinguished from either merely-subjective time or literal-wristwatch time).

§6. Summary Of Findings

Based on logical, ontological, and other relevant considerations specified above, it was concluded that in the very-long-run: (1) forward-directed time travel capacity is highly likely; and, (2) past-directed time travel capacity is likely. Four logically possible forward-directed, and four logically possible past-directed, types of (hypothetical) time machines were identified. Two different approaches (the "practical" and the "bi-temporal") were utilized in attempting to characterize the meaning of time travel. It apparently turns out that the concept of "embedded-subjective time" (i.e. the embedded-temporality of the human time-traveler, as distinguished from either merely-subjective time or literal-wristwatch time) is especially helpful in characterizing whether time travel did or did not occur in a particular circumstance.

§7. Two Closing Quotations

> What is time, then?
> As the future, it is possibility;
> as the past, it is the bond of fidelity;
> as the present, it is decision.
>   – Karl Jaspers[7]

> Mankind lies groaning, ... deciding if
> they want merely to live, or intend to
> make just the extra effort required for
> fulfilling ... the essential function
> of the universe, which is a machine
> for the making of gods.
>   – Henri Bergson[8]

Acknowledgements

I would like to thank Bob Ettinger, Steve Luper, and Mike Perry, and Ser-Min Shei and his philosophy department at National Chung Cheng University (Taiwan), for their assistance.

Bibliography

Bergson, Henri, 1932. *The Two Sources Of Morality And Religion*. [translated by R. Ashley Audra and Cloudesley Brereton with the assistance of W. Horsfall Carter]. (University Of Notre Dame Press). 1932, 1935, 1977.

Bostrom, Nick, 2005. [Dr. Bostrom's philosophy papers and more] <http://www.nickbostrom.com>. [retrieved 2005].

Broderick, Damien, 1999. *The Last Mortal Generation*. (New Holland Publishers).

Catterson, Troy T., 2003. "Letting The Dead Bury Their Own Dead: A Reply To Palle Yourgrau". Pages 413-426 In: Tandy, Charles (editor). *Death And Anti-Death, Volume 1*. (Ria University Press). Also at: <http://cetandy.tripod.com/ troycatterson/index.html>.

Chaisson, Eric J., 2001. *Cosmic Evolution: The Rise Of Complexity In Nature*. (Harvard University Press).

Deutsch, David, 1997. *The Fabric Of Reality: The Science Of Parallel Universes – And Its Implications*. (Penguin).

Drexler, K. Eric, 1987. *Engines of Creation*. (Anchor Press/ Doubleday).

Ettinger, Robert C. W., 2005. *The Prospect Of Immortality*. (Ria University Press). 1962, 1964, 2005.

Ettinger, Robert, 2006. "The Principles Of Experience". <http://www.cryonics.org/principles.html>. [retrieved on 6 September 2006].

Feinberg, Gerald, 1968. *The Prometheus Project*. (Doubleday).

Fromm, Erich, 1941. *Escape From Freedom*. (Rinehart & Co.).

Globus, Al, 2004. *Space Settlements*. <http://www.nas.nasa.gov/Services/Education/SpaceSettlemement/>. 2004. [retrieved 2004].

Hawking, S. W., 1992. "Chronology Protection Conjecture". *Physical Review D* (Volume 46, Issue 2: Pages 603-611). [15 July 1992].

Heinlein, Robert A., 1957. *The Door Into Summer*. (Doubleday). [Serialized 1956].

Jaspers, Karl, 1932. *Philosophy: Volume 1*. [translated by E. B. Ashton]. (University Of Chicago Press). 1932, 1969.

Kaku, Michio, 2004. *Parallel Worlds*. (Doubleday).

Lockwood, Michael, 2005. *The Labyrinth Of Time*. (Oxford University Press).

Lucas, J. R., 1989. *The Future*. (Basil Blackwell).

Mellor, D. H., 1998. *Real Time II*. (Routledge).

Merkle, Ralph, 1992. "The Technical Feasibility Of Cryonics". *Medical Hypotheses*. 1992. (Volume 39: Pages 6-16).

Naam, Ramez, 2004. *More Than Human*. (Random House).

O'Neill, Gerard K., 1975. [1975 Interview]. <http://lifesci3.arc.nasa.gov/SpaceSettlement/CoEvolutionBook/Interview.HTML>. [retrieved 2004].

O'Neill, Gerard K., 2000. *The High Frontier: Human Colonies In Space*. (Morrow). 1977. [A year 2000 reprint (from Collectors Guide Publishing, Inc.) contains updated information and a CD-ROM.]

Oppenheimer, J. Robert, 1984. *Uncommon Sense*. (Birkhauser).

Perry, R. Michael, 2000. *Forever For All: Moral Philosophy, Cryonics, And The Scientific Prospects For Immortality*. (Universal Publishers).

Pickover, Clifford A., 1998. *Time: A Traveler's Guide*. (Oxford University Press).

Polak, Fred, 1973. *The Image Of The Future*. [Boulding, Elise, translator.] (Jossey-Bass). 1953, 1973.

Popper, K. R., 1956 & 1991. *The Open Universe: An Argument For Indeterminism.* (Routledge).

Regis, Ed, 1990. *Great Mambo Chicken And The Transhuman Condition.* (Addison-Wesley).

Rosin, Carol, 2004. [The Institute For Cooperation In Space (Website)]. <http://www.peaceinspace.com>. [retrieved 2004].

Segall, Paul, 1989. *Living Longer, Growing Younger.* (Times Books).

Sen, Amartya, 2000. *Development As Freedom.* (Anchor Books). 1999, 2000.

Tandy, Charles, 2005. "The Emulation Argument: A Modification Of Bostrom's Simulation Argument". Pages 279-300 In: Tandy, Charles (editor). *Death And Anti-Death, Volume 3.* (Ria University Press).

Tipler, Frank J., 1994. *The Physics Of Immortality.* (Doubleday).

Tipler, Frank J., 2007 (forthcoming). *The Physics Of Christianity.* (Doubleday).

Vinge, Vernor, 1993. "The Coming Technological Singularity". *Whole Earth Review.* Winter issue. 1993.

Young, George, 1979. *Nikolai F. Fedorov: An Introduction.* (Nordland Publishing Company).

Endnotes

1. Heinlein 1957.

2. Hawking 1992. Also See: <http://www.hawking.org.uk/lectures/warps3.html>; And Also See: <http://arxiv.org/abs/ gr-qc/9703024>.

3. Compare: Ettinger 2006.

4. Popper 1956 & 1991, p. 61.

5. Lucas 1989, p. 1.

6. It may be remarked that the distinction between AOK Superfast Rocketry and EGC Superfast Rocketry is not relevant within the context of this paper; if so, this would yield 7, not 8, types of time travel. Or it may be remarked that the (here-relevant) distinction between AOK Superfast Rocketry and EGC Superfast Rocketry should also be applied to the other types of time travel; if so, this would yield 14, not 8, types of time travel.

7. Jaspers 1932, p. 57.

8. Bergson 1932, p. 317.

Chapter 12
Teleological Causes and the Possibilities of Personhood

"Teleological Causes And The Possibilities Of Personhood" was first published in 2007 and is here reprinted by permission.

> For the history that I require and design, special care is to be taken that it be of wide range and made to the measure of the universe. For the world is not to be narrowed till it will go into the understanding (which has been done hitherto), but the understanding is to be expanded and opened until it can take in the image of the world.
> – Francis Bacon (1620)

Today, in our age of the global village, historians try to write global histories, as distinguished from geographically limited or clannishly insular accounts. Some even try to do "big history" or more, seeking to go beyond fixation on a single species (humans) and to include all geographical regions – perhaps even attempting to include all temporal periods in the long history of the universe. Indeed, these days some scientists even speak seriously, if speculatively, of a possible "multiverse" in which our entire universe is infinitesimally tiny by comparison.

It does seem that our image of the universe or multiverse grows in size and profundity as we learn more and more. The world presented in Dante's *Divine Comedy* (early 14th century) induces claustrophobia in today's physicist. Yet Dante thought he was presenting everything – not only everything in this world but everything in Heaven, Hell, and Purgatory as well.

In this paper I will assume that the whole of Everything is indeed immense and profound. I will suppose that something like a "many-worlds" or "multiverse" or "meta-verse" view of things is correct – without defending it. In part one ("The Many-Multiverses (M-M) Model"), I postulate certain metaphysical and existential assumptions which serve to model the larger world (or multiverse of multiverses) in which we are said to be embedded. In part two ("The Possibilities Of Personhood"), I describe the general sorts of individual physical (time-space) entities that exist in this region of this universe or that may exist in other or future regions of the multiverse of multiverses. (An anti-speciesist stance results.) Finally, in part three ("Super-Persons As Teleological Causes"), I take for granted the context outlined in parts one and two so as to consider ethical-teleological issues related to the emergence of super-persons (quasigods) from persons.

Part One: The Many-Multiverses (M-M) Model

> Mankind lies groaning, ... deciding if they want merely to live, or intend to make just the extra effort required for fulfilling ... the essential function of the universe, which is a machine for the making of gods.
> – Henri Bergson (1932)

A number of "serious-but-speculative" theories of the multiverse have been offered by today's scientists and philosophers. Moreover, each approach typically produces its own competing versions when specifics or details to the theory are attempted. My own approach below (I call it "the Many-Multiverses Model") has certainly been influenced by the "Matrix" theory of Nick Bostrom (2003).

One of the strengths of my model is that it is in some sense compatible with a variety of multi-verse and even uni-verse theories. You may identify my paradigm with either a multiverse or a universe – but I personally think of it as a "many-multiverses" or "meta-multiverses" model (a multiverse of multiverses of multiverses of ...). My positing of "many-multiverses" assumes you are open to the following argument or considerations:

↓   ↓

Precisely one of the following three hypotheses must be true:

Hypothesis One: Advanced life or reflective-personhood (sapient-personhood) never reaches a super-personhood (quasigodhood) stage. (Example: Persons will always become extinct before reaching a super-person or quasigod stage of development.)

Hypothesis Two: Advanced life or reflective-personhood (sapient-personhood) reaches a super-personhood (quasigodhood) stage, but never produces (engenders) new universes.

Hypothesis Three: Advanced life or reflective-personhood (sapient-personhood) reaches a super-personhood (quasigodhood) stage and produces (engenders) new universes.

And also the following reasoning is reliable:

Every existing universe is God engendered, Nature engendered, and/or Super-person engendered; We live in an existing universe; **Therefore**: We are now living in a God engendered universe, a Nature engendered universe, and/or a Super-person engendered universe. (Note that the term "and/or" -- rather than the term "or" -- is used here.)

↑ ↑

"We are now living in a God engendered universe, a Nature engendered universe, and/or a Super-person engendered universe." This seems to cover the relevant (three) sets of possible universe-engendering entities (God; Nature; Super-persons), including the possibility of universes being stacked on top of each other. This means that, in the broad view, it is possible to live in a universe that was engendered at, say, a God level, then a Nature level, then a Super-person (Quasigod) level; indeed, it can get more complicated, much more complicated, than this. This conclusion appears to clearly and correctly state the metaphysical alternatives in a manner conducive to stimulating further philosophic insight, to encouraging a reasonable basis for common dialogue, and to attempting a global advancement of learning.

Perhaps this argument or presentation or conclusion should persuade us to take seriously both the "Super-person" ("Quasigod") Scenario and the "Extinction" ("Doomsday") Scenario. Thus it is easy to believe that there may be many (and many levels of) universes beyond the universe in which we happen to find ourselves. Moreover, one may argue that the idea of the existence of many different universes produced by many different Super-persons (Quasigods), if taken seriously, supplies some reasoned consolation to would-be theists who have difficulty with the problem of evil -- not with the problem of evil free agents, but with the problem of an evil natural world, such as animal and human morbidity, pain, mortality, and ignorance. (A theist might now say: God does not create evil but God does allow each person capable of moral reflection, including each super-person, to exercise free-will for good and evil.)

One may complain that my many-multiverses ("m-m") model should (attempt to) address (if it can) the issue of infinity with reference to the past. For example, it is possible that there are not only an infinite number of future levels of universes -- but also an infinite number of past levels of universes as well. Yet another possibility: Our universe, or a previous universe related to the engendering of our universe, has always existed. In these two scenarios about the infinite past we can see that there would be no "First Cause" as traditionally or typically conceived. This writer is open to further education and insight on the matter, but is presently inclined to believe that the possibility of an "infinite past" is not contrary to his m-m model. For example, we can think of "the infinite past" (if there is such a thing) as "the way things (Naturally or Divinely) are". One may nevertheless ask as to the proper meaning of "engendered" as used above (or whether a better word could be chosen). This writer is open to suggestions.

After recently reading Kierkegaard (1847) and Penrose (1994), it occurred to this writer that the three metaphysical alternatives arrived at in the conclusion above are not really so new: (1) "**God**": "God" has some similarity to Kierkegaard's "the eternal" and to Penrose's "Platonic world". (2) "**Nature**": "Nature" has some similarity to Kierkegaard's "the temporal" and to Penrose's "physical world". (3) "**Super-persons**": Religious folks have often talked about "the divine spark within" -- with personhood (Kierkegaard's "the single individual") serving as a bridge between the physical and the divine; for Penrose, the "mental world" (identified with understanding and insight) is the third of his three real worlds that ultimately compose one world; and my m-m model refers to "reflective-personhood" as well as to "super-personhood" ("quasigodhood"). When we speak of personhood or super-personhood (quasigodhood), we are not necessarily referring specifically to a biological entity. The m-m model as presently conceived is agnostic about the issue of "substrate independence".

We can think in terms of something like Super-personhood (Quasigodhood) or Reflective-personhood or Personhood or the mental world as potentially universe-engendering. We can think in terms of something like Nature or the temporal or the physical world as potentially universe-engendering. And we can think in terms of something like God or the eternal or the Platonic world as potentially universe-engendering.

Thus the m-m model would seem to cover the three (non-exclusive) metaphysical alternatives with respect to universe-making (God; Nature; and, Super-personhood or Quasigodhood). In addition, the m-m model would seem to cover **our** exclusive existential alternatives:
(1) We do not reach Super-personhood (Quasigodhood).
(2) We reach Super-personhood (Quasigodhood) but do not engender universes.
(3) We reach Super-personhood (Quasigodhood) and engender universes.

Part Two: The Possibilities Of Personhood

> [A person is] a type of entity such that *both* predicates ascribing states of consciousness *and* predicates ascribing corporeal characteristics ... are equally applicable to a single individual of that single type.
> – P. F. Strawson (1959)

We do not know which one of the three existential alternatives will be **our** actual future. (Arguably we have some control over our future.) We do not know if **our** particular universe was proximately engendered by God or by Nature or by a Quasigod. There is much we do not know or do not yet know; nevertheless we have to act in the present. Yet -- as we ponder the vastness (infinity? infinity of infinities?) of a multiverse of multiverses, it seems reasonable to believe that at least **some** persons in **some** universes will reach super-personhood (quasigodhood) regardless of **our** outcome.

Beyond this, it does seem that the entities or forms we (partially and fallibly) discover in mathematics and logic are real, even eternally (a-temporally) real. It also seems that moral values must be real, even eternally (a-temporally) real. For example, it seems to make a good deal of sense to talk about what is authentically (objectively or really or truly) in one's (or our) best interest even if we are unsure what exactly that "best interest" is in a particular situation. Such considerations as these seem to tell us that something like God or the eternal or the Platonic world is real rather than hypothetical.

I (this writer) first read George Orwell's novel *1984* many years ago (many years before the year 1984). I read the novel because I thought it was science fiction -- but found it to be much more. By the end of the dystopia, even Winston Smith has been thoroughly brainwashed. If his boss holds up his hands saying he has twelve fingers, Winston **actually sees** twelve fingers. Winston **clearly remembers** past events -- but they are events that never **really** took place. This dramatic ending to the novel engrained in me both a sense of the difficulty of uncovering past truths and a belief in the actual existence of the past. (For a detailed philosophic argument that the past will necessarily always exist, see Catterson 2003.)

Once I do X instead of Y, X will **always** be the case. It is impossible for the past to be annihilated; the past necessarily forever continues to exist. It is not just a linguistic convention when we sometimes speak of the past as presently existing. It is not immediately obvious to what extent we or others (say, Quasigods) will ever be able to access such existences we call past. But in principle it does not seem to be altogether impossible; e.g. perhaps our universe was engendered (and is "recorded") by a Quasigod (Super-person). (And, according to Tandy 2007, past-directed time-travel capacity is "likely" in the very-long-run.)

John Rawls (1971) is famous for his "veil of ignorance" thought experiment. To probe more deeply into the kind of society we should want, we can imagine we are constructing the moral-political basis for such a society under a veil of ignorance. In our mutual contracting with each other we do not know our position in the imaginary or new society -- we may be socially-economically high or low, we are ignorant of our religious-philosophical beliefs, etc. Since we are ignorant of our own special interests and status and roles in the society, we will want a society that treats everyone fairly.

Previously we spoke of our ignorance of the metaphysical place of our own little world in the total scheme of things. This actual situation of ignorance can function in a way analogous to the Rawlsian hypothetical situation of ignorance, but with a more potent practical force. "Treat others as you would have others treat you." "Respect all persons, past-present-future." It seems that whether we are relatively rich or poor, or whether we are transmortal quasigods or mortal humans, the golden rule applies to all persons with the capacity to reflect on matters of good and evil.

Indeed, to what extent those of our world and those of other worlds can learn the great golden lesson will affect the quality of all our lives. Likewise, we can consider the parallel "golden rule" functions of ending poverty, canceling doomsday, and progressing toward Quasigodhood (Super-personhood). We can consider the parallel functions of achieving victories over morbidity, pain, mortality, and ignorance.

At a theoretical-intellectual level we can distinguish the basic kinds of individual time-space (physical) entities as follows:

> **Non-Persons** ("corporeal characteristics" but no capacity for "states of consciousness")
    1. Objects. Objects have no capacity for consciousness.

> **Persons** (both "corporeal characteristics" and capacity for "states of consciousness")
    2. Mere-Persons. Mere-persons have the capacity of consciousness but not the capacity of moral reflection.
    3. Sapient-Persons. Sapient-persons have the capacity of consciousness and including the capacity of moral reflection.

Note that in our consideration of to what extent an entity is (1) an object; (2) a mere-person; or, (3) a sapient-person – we should obviously NOT base the evaluation on the SPECIES to which the being is said to belong. For example, some individual members of the human species (newborn humans; adult humans continuously severely mentally impaired from birth) do NOT belong in category (3) above. For example, some individual members of non-human species (some individual non-human animals) DO belong in category (3) above. The "real-life" boundaries between the species are NOT sharp; in addition, the "real-life" boundaries between the three categories above are NOT sharp. The present "anti-speciesist" paragraph should be kept firmly in mind when correcting or correctly-interpreting the remainder of the paper below.

In order to gain further ethical-teleological insight into speciesism and personhood, I will now proceed to articulate some additional relevant parameters and distinctions. Since we know little of our own universe and even less of other universes (and even less of the whole multiverse of multiverses), we will have to proceed as best we can – you and me, here and now. I will begin with what today's scientists tell us about the history of our universe, and I will engage my moral imagination to encompass additional relevant futuristic considerations:

(1) With the "big-bang" of our own universe's time-space came **energy** -- and phase-states.
(2) Then came **matter** -- and (atomic/molecular) valences.
(3) Then life (**vegetation**) -- and creodes (tropisms, genetics, and/or algorithms) (See Waddington 1967: "creodes" teleonomically explain why a baby rose-flower becomes an adult rose-flower instead of an adult daisy-flower; why a kitten becomes a cat instead of a dog; and why an adult human becomes an aged-dying human instead of a human/non-human child.)
(4) Then (after plants) comes **mere-persons** (animals) -- and feelings-desires (hedonics: pleasure and pain).
(5) Then **sapient-persons** (humans) -- and second-order (higher-order) desires-feelings (free-will reflection on good and evil).
(6) Then **super-persons** (quasigods) -- and good-evil dialogue/deliberation to renovate (and/or engender) one or more universes.
(7) And finally (?) comes **Ultimate-Personhood** and realization of the Good (?). (For more information about Ultimate-Personhood, see Tandy 2002.)

Here is another way to present it in outline form:

> **Non-Teleological Entities** [**and** each non-person's characteristic Modus Operandi]
   1. "**Primal-Energy**" [**and** phase-states]
   2. "**Basic-Matter**" [**and** (atomic/molecular) valences]
   3. "**Teleonomic-Systems**" [**and** creodes (tropisms, genetics, and/or algorithms)]

> **Teleological Beings** [**and** each person's characteristic Modus Operandi]
   4. "**Mere-Persons**" [**and** desire or feelings-desires (hedonics: pleasure and pain)]
   5. "**Sapient-Persons**" [**and** reflection or second-order (higher-order) desires-feelings (free-will reflection on good and evil)]
   6. "**Super-Persons**" (or "Quasigods") – either "**Godly** Super-Persons/Quasigods" or "**Ungodly** Super-Persons/Quasigods" or perhaps somewhere in-between good/godly and evil/ungodly)

[**and** good-evil dialogue/deliberation to renovate (and/or engender) one or more universes]
7. "**Ultimate-Personhood**" [**and** realization of the Good (?) and/or re-presentation of time-space **facts** and eternal truths (logical-mathematical **forms** and aesthetic-moral **values**) (?)]

Part Three: Super-Persons As Teleological Causes

> In the long run truth wins the race
> with falsehood and error – and it
> wins because it is truth.
> – E. A. Burtt (1965)

Almost all of reality is located in the future. With reference to our known universe, it seems that non-persons (non-teleological forces or causes) have dominated the past and that persons (teleological forces or causes) may well dominate the future. But today's persons are hugely influenced by the past from which they emerged. However, a future person may have the capacity to reinvent oneself, to restructure one's own non-teleological (energy, matter, teleonomic) systems and also one's own hedonic system to conform to the results of one's own moral reflection or immoral choice.

Restructuring the energy-system of one's own body might involve advanced subatomic technology as well as insight into reasonable expectations. Restructuring the matter-system, the teleonomic-system, and the hedonic-system of one's own body might involve advanced molecular (nano) technology as well as insight into reasonable expectations. It is of course conceivable that modifying one system might have unknown consequences for the other systems.

I'm not sure we know enough about energy or subatomic technology to yet offer responsible advice about the restructuring of the energy-system of one's own body. However we do have some beginner's insight into the advanced molecular (nano) technology of the future. We may want to begin with modest modifications to our bodies as we gradually learn more. "A little knowledge is a dangerous thing."

I will make a few brief remarks related to the teleonomic-system and the hedonic-system. The teleonomic and hedonic systems of today's person are structured based on the non-teleological past. This suggests that great changes to these systems are in the long run to be preferred so as to enhance the lives of persons.

Some may believe that a teleonomic system (whether of a rose-flower or of a human) is teleological because it seems to exhibit purposefulness and is goal-oriented. But in fact the teleonomic-system as such is NOT conscious and is the result of evolutionary adaptation. Although there may be good practical reasons for taking a cautious approach to its modification, from a moral-teleological point of view its improvement is imperative. Thus in a thought experiment (rather different from our actual world context, or so I believe) we can imagine a world context in which, as a practical matter, there may be good reasons for not extending the healthy lifespan of persons from 50 years to 500 years. In the world in which we actually live, however, my sense is that such so-called reasons are not really very good reasons – we are biased by confusing teleonomy with teleology.

Likewise, many fail to see that our hedonic-system (of pleasure and pain) is also based on the past and should be modified with advice from our system of moral reflection. Pleasure and pain, given advanced future technology, could presumably be structured in a wide variety of different ways. (To be sure, a variety of hedonic-systems already exist.) We could structure it so that good behavior is painful and bad behavior is pleasurable. Alternatively, we could structure it so that philosophic reflection and moral behavior are the most pleasurable of pleasures. The point is that "having fun" is not the highest value, but with future technology we will presumably be able to restructure our system of pleasure and pain to make it more ethical-teleological.

As our universe became more complex, moral consciousness eventually appeared. Now moral consciousness must learn to unbias or free itself from the teleonomic and hedonic systems of old in order to renovate the blind universe. The blade of grass is digesting the dirt, while the insect is eating the blade of grass, while the mammal is devouring the insect. The mammal, caught in a metal trap, sees the human hunter approaching. The blind universe has cruelly set animal against animal – and humans against mortality.

Here are some of the presumed capacities of Super-persons as they renovate (well or poorly) the blind universe:

- Use of free-will and great power to pursue wisdom, to learn self-respect, and to respect all persons, past-present-future.
- Insure that no animal kills another animal. This includes both non-human animals and human animals.
- Insure that no reflective-person must die.
- Insure that no person must experience unwanted serious pain or hardship.

With Super-personhood (Quasigodhood), we may have the ability to run ancestor history emulations (via time travel or otherwise). R. Michael Perry (2005) has remarked that it would seem to be immoral to run such ancestor histories – real persons would experience real pains and evils. Instead, as Perry advocates, the golden rule would charge us with the duty to revive our ancestors – the scientific resurrection of all dead persons in the history of the multiverse of all multiverses.

Bibliography

Bacon, Francis (1620). See: Eiseley (1973).

Bergson, Henri (1932). *The Two Sources Of Morality And Religion*. Translated by R. Ashley Audra and Cloudesley Brereton with the assistance of W. Horsfall Carter. University Of Notre Dame Press: Notre Dame. (1932, 1935, 1977). Page 317.

Bostrom, Nick (2003). "Are You Living In A Computer Simulation?" *Philosophical Quarterly* 53(211) [2003]: Pages 243-255. Also see a Nick Bostrom website: <http://www.simulation-argument.com>.

Burtt, E. A. (1965). *In Search Of Philosophic Understanding*. New American Library: New York. (1967 Edition). Page 5.

Catterson, Troy T. (2003). "Letting The Dead Bury Their Own Dead: A Reply To Palle Yourgrau" Pages 413-426 In: Tandy, C. [Editor] (2003). *Death And Anti-Death, Volume 1*. Ria University Press: Palo Alto, CA.

Eiseley, Loren (1973). *The Man Who Saw Through Time*. Charles Scribner's Sons: New York. (1961, 1962, 1964, 1973). Francis Bacon quoted on page 9.

Ettinger, R. C. W. (2002). "Youniverse" Pages 237-272 In: Tandy, C. And Stroud, S. R. [Editors] (2002). *The Philosophy Of Robert Ettinger*. Ria University Press: Palo Alto, CA.

Ettinger, R. C. W. (2004). "To Be Or Not To Be: The Zombie In The Computer" Pages 311-338 In: Tandy, C. [Editor] (2004). *Death And Anti-Death, Volume 2*. Ria University Press: Palo Alto, CA.

Kierkegaard, Soren (1847). *Works Of Love*. Translated by H. Hong and E. Hong. Harper & Row: New York. (1962, 1964).

Lepore, E. And Van Gulick, R. [Editors] (1991). *John Searle And His Critics*. Basil Blackwell: Oxford.

Penrose, Roger (1994). *Shadows Of The Mind*. Oxford University Press: New York.

Perry, R. Michael (2000). *Forever For All: Moral Philosophy, Cryonics, And The Scientific Prospects For Immortality*. Universal Publishers: Parkland, FL.

Perry, R. Michael (2005). Personal Communication From R. Michael Perry To Charles Tandy (21 August 2005).

Rawls, John (1971). *A Theory Of Justice*. The Belknap Press Of Harvard University Press: Cambridge, MA. Revised Edition, 1999. (Original Edition, 1971).

Searle, J. (1980). "Minds, Brains And Programs." *The Behavioural And Brain Sciences* 3 [1980]: Pages 417-57.

Searle, J. (1984). *Minds, Brains And Science*. Harvard University Press: Cambridge, MA.

Strawson, P. F. (1959). *Individuals: An Essay In Descriptive Metaphysics*. Routledge: London. (1959, 1964, 1990). Page 102.

Tandy, Charles (2002). "Toward A New Theory Of Personhood" Pages 157-188 In: Tandy, C. [Editor] (2002). *The Philosophy Of Robert Ettinger*. Ria University Press: Palo Alto, CA.

Tandy, Charles (2003). "N. F. Fedorov And The Common Task: A 21st Century Reexamination" Pages 29-46 In: Tandy, C. [Editor] (2003). *Death And Anti-Death, Volume 1*. Ria University Press: Palo Alto, CA.

Tandy, Charles (2007). "Types Of Time Machines And Practical Time Travel" ***Journal Of Futures Studies*** 11(3) [February 2007]: Pages 79-90.

Waddington, C. H. (1967). ***The Ethical Animal.*** University Of Chicago Press: Chicago. Page 82.

Chapter 13
Terrestrial Peoples, Extraterrestrial Persons

"Terrestrial Peoples, Extraterrestrial Persons" was first published in 2007 and is here reprinted by permission.

Herewith I join Gary Hart in the invisible college or visible club to support the Daalder and Lindsay proposal for a Concert of Democracies. [1,2] (The article by Ivo Daalder and James Lindsay is entitled "Democracies of the World, Unite".) [3] It is an idea whose time has come. Now is the time for its implementation. The Concert or Union may be expected to provide a number of benefits to world betterment, including: (1) Strengthening and expanding the positive relationship zone of peace and peaceful activities among democratic (or Rawlsian well-ordered) societies; and, (2) Weakening the negative temptations of democratic (or Rawlsian well-ordered) societies, such as (A) crusading imperialism; (B) imprudent appeasement; and, (C) moralistic isolationism.

Professional philosophers (of which I am one) often talk an idea to its death. Usually this is a good thing. But sometimes history opens up a new niche which is too good to pass up if we seek world betterment. In Part One below I raise questions about the Daalder and Lindsay proposal for a Concert of Democracies, but my intent is not that of criticizing it to its death. My intent, rather, is to hasten the proposal's rapid adoption by the liberal democracies of the world. If my questions may hinder the proposal's approval, then we should first firmly establish the needed new organization (the Concert) and only then raise the matter of possible modifications. I say this because I do not want anyone to misinterpret my intentions in Part One below. On the other hand, Part Two below should be taken at face value. I propose a second organization I believe also desirable and feasible for implementation now. Each organization would substantially improve the world – but I believe implementing both proposals in concert would have a synergistic effect, especially with reference to improving the far future.

1. Part One

Immanuel Kant's *Perpetual Peace*, published in 1795, is a remarkable piece of social science foresight. [4] In 1795, few republics existed or no liberal democracies existed (e.g. consider civil rights issues related to slavery and women). Kant argued for republicanism and for an expanding concert of peaceful republics. He believed this approach (and not the universal membership approach) would eventually lead (in the 21st century?) to a global stable peace.

With reference to the Daalder and Lindsay proposal, we raise the following question: Do we want the new organization to be a concert of (1) liberal democratic states; (2) well-ordered states; (3) liberal democratic peoples; or, (4) well-ordered peoples? John Rawls opts for well-ordered peoples. He thinks we should think in terms of peoples rather than (Westphalian) states if we want to be open to the future instead of wedded to the past. Likewise he thinks we should be open and humble even in the face of his *Theory of Justice* (liberal democracy as the end of history). (I believe it was Soren Kierkegaard who once wrote that Hegel might be the greatest philosopher, except that Hegel forgot to say "this is one individual's opinion.")

In section 8.1 of *The Law of Peoples*, John Rawls identifies five types of societies: (1) liberal peoples; (2) decent peoples; (3) outlaw states; (4) societies burdened by unfavorable conditions; and, (5) benevolent absolutisms. Liberal peoples and decent peoples, considered together, are referred to by Rawls as "well-ordered" peoples. In sections 4.1, 4.2, and 12.1 of his *Law*, Rawls attempts to specify the general or basic requirements for a well-ordered society. In principle, a non-democracy (e.g. a theocracy) might be able to qualify as a well-ordered society. Rawls points out, however, that at this precise moment in history, no non-democracies would in fact qualify for membership in the Concert or Union. And even in principle it is impossible for an illiberal democracy to qualify for membership.

I add a personal note here. I have been living in Taiwan for a number of years – so I am biased toward the peoples (instead of states) approach. Taiwan (which is a liberal democratic people) could be a founding member of a concert of (well-ordered) peoples even though it is not a member of the UN. China (which is not well-ordered) could not be a founding member of a concert of well-ordered peoples even though it is one of the Big Five (permanent/veto) members of the UN.

2. Part Two

Above, I endorsed the proposal for a "Concert of Democracies" (CD) or "Union of Peoples Well-ordered" (UP). Now I present an additional proposal endorsing a "Treaty Organization Acting for a Better

Cosmos" (TO or TO-ABC). [5] I believe that both the CD/UP and TO proposals are desirable and feasible for today's world. These two Concerts, acting more or less in concert, may have historically unusual abilities to transmute our civilization of outmoded States into a transcivilization of authentic Communities.

The CD/UP idea pioneered by Kant and Rawls now seems obvious to me, thanks to Daalder and Lindsay. Another idea whose time has come (the TO) may be less obvious until some folks like Daalder and Lindsay come along to show us the way – to present the TO idea so that it suddenly seems obvious. I should expect that some folks will come along with an improved "breakthrough" presentation to make the TO proposal more obvious.

Below I will attempt such a presentation. I will begin by having you consider the capacity of future technology. After that, we will consider what we can and should do today (the TO proposal) if we are to guide our technology so as to take us from bad to better instead of from bad to worse.

As we consider life and technology in a transcivilized future, let me ask you these questions: Is there any doubt but that in the long run many of our offspring will be permanently living and working somewhere in the universe other than on planet Earth? Is there any doubt but that in the very long run almost all of our offspring will be born and permanently living somewhere in the cosmos other than in our Solar System?

The astounding capacity of future technology can be glimpsed at by taking a non-controversial look at the future of O'Neill Habitats and of Drexler Technology (and their eventual melding together). [6,7] (Indeed, Dr. Drexler was a student of Dr. O'Neill.) I say non-controversial because the controversy in each case is over when, not if. For present purposes we can overcome this dispute by simply talking non-controversially about these kinds of capacities in the far future (thus bypassing timeline predictions of near or far).

2.1 O'Neill Habitats
(Extraterrestrial Green-habitat Communities: "EGCs")

The fact that Earthlings presently exist together in a single biosphere global village is a rather absurd position to be in if we seek to prevent doomsday and promote flourishing. If something catastrophic happens to Earth's biosphere, then something catastrophic happens to all Earthlings.

It is not wise to put all of humanity's eggs (futures) into one basket (biosphere).

Extraterrestrial Green-habitat Communities ("EGCs") should not be confused with space stations. Some argue that if we had chosen to do so, we could have started building EGCs using the "merely super" technology of the 20th century. Indeed, the famous 20th century physicist Gerard K. O'Neill designed such EGCs for the purpose of late 20th century construction. Such EGCs would provide a "green-friendly" environment for humans, animals, and plants superior to the problematic habitats we identify with Earth and other planets.

Eventually millions of persons in a single EG Community are possible. The EGCs would be self-sufficient and could reproduce other EG habitats in extraterrestrial space at a geometric rate. Accordingly, there is "unlimited free land" in extraterrestrial space -- with a higher quality of life than is possible on the surface of a planet.

EG Communities can be built from extraterrestrial resources mined from asteroids, planets, or moons. Rotation of the large and spacious greenhouse habitat provides simulated gravity for the people and plants living on the inner surface. Adjustable mirrors provide energy from the sun and simulation of day and night. Sooner or later, the following would be feasible for EGCs:

- "Unlimited energy" from the sun.
  (The sun never sets in space.)
- Control of daily weather and sunlight.
- Self-sufficient EGCs.
- Expansion of self-sufficient EGCs at a geometric rate.
- "Unlimited free land" via EGCs.
  (Needed raw materials from asteroids are abundant.)

The following metaphorical insights have been widely quoted by EGC experts: "The Earth was our cradle, but we will not live in the cradle forever." "Space habitats [EGCs] are the children of Mother Earth." According to Carl Sagan, our long-term survival is a matter of "spaceflight or extinction." According to the "mass extinction" article in the latest (6th) edition of *The Columbia Encyclopedia*, "The extinctions, however, did not conform to the usual evolutionary rules regarding who survives; the only factor that appears to have improved a family of organisms' chance of survival was widespread geographic colonization." [8]

2.2 Drexler Technology
(Molecular Nano Technology: "MNT")

Molecular nanotechnology (MNT) is not required for the development of EGCs (independent, self-sufficient, self-replicating biospheres in extraterrestrial space). Advanced MNT certainly will greatly enhance EGC capacities, however. But it will do far more.

Micro-technology and 20th century nano-technology were pioneered by Japan and the USA; such technology explains in part America's victory in their cold war "space race" with the Soviet Union. With micro-technology we made things smaller and lighter than ever before. Toward the end of the 20th century we went beyond mere micro-technology to what some called nano-technology (a mini-object produced or its mini-parts might no longer be visible to the naked eye even with the aid of an optical microscope).

Although we can think in terms of making things smaller and smaller or in terms of an evolution from micro-technology to nano-technology, molecular nanotechnology (MNT) in its advanced form will approach manufacturing or production of objects, circuits, parts, foods, computers, robots, software and devices using a radically different strategy. Until the 21st century our strategy had been to make things smaller. But the strategy of 21st century MNT is to make things larger (larger than molecules and atoms) by assembling molecules and atoms to any configuration permitted by the laws of nature. MNT is the way nature does things, building from the bottom up. MNT in nature gives as all sorts of plants and animals; we throw a seed in the ground and latter find a watermelon there; MNT produces a human infant in only nine months.

Human-designed MNT will eventually produce nano-size computers, nano-size factories, and molecular-repair nano-robots. MNT in its advanced form will have profound biological and biogenetic implications – e.g. the capacity to defeat all disease, including age-related death and disability. (Even accidents may become less frequent -- but it seems to be in the nature of some accidents that they are not predictable in advance.) Technology to clean up toxic waste dumps, and the widespread development of inexpensive non-polluting ("carbon-neutral") advanced technology, becomes feasible. Meat-eaters will not have to hurt or kill animals in order to eat meat; MNT will eventually be able to manufacture (actual) meat to the specification of meat-eaters. (But even today where I live in Taiwan, some popular restaurants are quite good in offering vegetarian foods that look and taste like meat, but actually contain no meat.)

2.3 Treaty Organization Acting for a Better Cosmos ("TO" or "TO-ABC")

We may not know the actual or secret (classified) policies of the USA and others with respect to extraterrestrial space. It is nevertheless true that over 100 nations (including all of the "major" ones) publicly claim to support the 1967 Outer Space Treaty. Article II of the treaty says in its entirety: "Outer space, including the moon and other celestial bodies, is not subject to national appropriation by claim of sovereignty, by means of use or occupation, or by any other means."

Historically one of the reasons Terrestrial civilizations engaged in wars against each other was to gain more territory, and the power and glory that came with empire. But the development of advanced EGCs will mean "unlimited free land" (freely available territory) and the realistic possibility of **intentional** (i.e. voluntary) communities for all persons. Instead of remaining in the community or culture of one's birth, one will be realistically free to experiment living in one kind of community or another. New kinds of cultures and communities will be enabled by the new extraterrestrial technology.

Eventually there will be many Extraterrestrials, few Terrestrials. We can understand the practical or special interests that might prevent us from banning weapons and their manufacture from today's Earth. Indeed, someday there might be analogous practical or special interests in extraterrestrial space unless we engage in foresight today to proactively ban weapons and their manufacture from extraterrestrial space.

On the one hand, our political interests today may constrain us in our present time and place. But, on the other hand, our political interests today may free us with respect to future times and places (e.g. our extraterrestrial future). What this means is that today we have a realistic prospect of proactively establishing the legal structure and enforcement powers needed for a world at stable peace in extraterrestrial space.

If we wait until later, we may not be so free to "do the right thing" and establish stable peace in extraterrestrial space. Extraterrestrial space is immense; it is all of the universe except for a single small planet and its atmosphere. Eventually it might even become feasible to extend stable peace to planet Earth and thus the entire universe.

I will spend most of the remainder of this article trying to "think through" what the structure of the Extraterrestrial Society should be like -- a structure we would contemplate, modify, and implement in the present **before** we live and develop special interests out there. Such "thinking through" to produce an Extraterrestrial Space Treaty (the TO proposal) might also help us better understand conflicts and their possible management on today's planet Earth. It is my belief that the suggested Extraterrestrial Space Treaty Organization (TO) will make a fine gift to our offspring and, by the way, help present Earthlings.

If we want a good world at stable peace (whether that world be Terrestrial Civilization or Extraterrestrial Transcivilization), it would seem we must be willing to unblinkingly face up to the following questions: Is stable peace possible if each person or each people is passionately convinced their worldview is basically good and correct -- and different worldviews are evil or bad or incorrect? If we could enforceably prevent each and every person from killing any person over a conflict (say, a conflict of worldviews), would we do so? If so, how would we resolve our conflicts?

Although I have freely borrowed ideas from others, I believe the political theory or scheme of moral-political notions I present below is original with me. One advantage we have in facing up to the difficult questions raised in the previous paragraph is that we can use our imaginations to futuristically view ourselves as Extraterrestrials living in intentional communities (EGCs). We can further assume that a political structure there and then exists that we describe as a good world at stable peace.

The Extraterrestrials of the future have transhuman liberties and technologies. The Terrestrials of the present do not have transhuman liberties and technologies. Yet humans today have the ability and perhaps the practical political will -- via the TO proposal -- to help insure the existence of transmortal transhumans and a good world at stable peace in extraterrestrial space (**almost** all of the universe) in which transcivilization will flourish. So we need to "work backward" to determine the provisions of the Treaty or Concert now under construction.

First of all, I will assume that it is a fact that if today's Terrestrials are to produce such a Concert or Treaty (including effective enforcement provisions), it will require agreement from a number of States/Peoples. I also assume that eventually a Treaty like this would have to be binding in the sense that the Treaty would have no expiration date. The first Treaty however might have an expiration date and might have few Parties (States/Peoples) to the agreement. As they consult with each other, with other countries, with philosophers, with scientists, with politicians, etc. they would gain important insights and experience helping them produce a second Treaty, this time with no expiration date but with many Parties to the agreement, this time also containing strong and effective enforcement provisions.

How many persons or states/peoples would accept or endorse a Space Treaty that effectively and enforceably bans weapons and their manufacture from extraterrestrial space? In this context (a good and practical legacy to our offspring), I should think we should be diligent enough to rally enough supporters. For example, this (the second?) Treaty might be signed originally by, say, twenty States/Peoples (including all or most of the "major" ones). **But the Treaty would be strongly effectively enforced by TO's Agency for a Better Cosmos (not by States/Peoples) against ALL and EVERYONE, whether or not they sign the Treaty.** Once in force, I would expect many others to sign on -- since the Treaty applies to them even if they do not sign it. Eventually the Treaty really would have to be strongly effectively enforced by the ABC against all and everyone, because eventually persons and communities (EGCs) will permanently settle in extraterrestrial space. Too, such a Treaty offers hope and inspiration to those of us of the present.

Okay, you may say, this is a reasonable enough start, but what other liberties, responsibilities, and political structures would be appropriate for the Extraterrestrial World? So far, what we presumably have is an Extraterrestrial World at stable peace. But what about conflicts and the plurality of deeply held religious and philosophic worldviews?

What seems to me both practical and fair in this context is to think in terms of an Extraterrestrial Society of Intentional Communities. There would be two sets of liberties and two sets of responsibilities (for "Extraterrestrial Society" and "Intentional Communities" respectively). Each person is free to found new (intentional) communities. Each Community would determine its own membership requirements. Each Community would have **its own** culture of liberties and responsibilities; a member would generally be free to leave the community. A mechanism or set of mechanisms would be established to insure that each member is fully and properly informed of their liberty to leave the (intentional) community. (I suppose some communities might still allow their members the possibility of experiencing physical pain -- but they would also allow a member to voluntarily leave their community. Too, I suppose banning animal cruelty and serious animal pain would be desirable and feasible.) Note that some ("hermit") communities would consist of only one person.

On old Terra, it was often difficult or impossible to leave one's community -- sometimes expulsion effectively meant the individual's death. The context of the Extraterrestrial Society of Intentional Communities is radically different. The individual transhuman person would be transmortal, whereas on old Terra it was often the community or society (not the human individual) that was seen as transmortal.

So at the level of the **Society** (of Communities) we have: (1) **Peace**: Weapons, weapons-making, and violence (including animal cruelty and serious animal pain) are strongly effectively enforceably banned; and, (2) **Freedom**: Every individual person is fully aware of and fully informed of their general liberty to leave their community. This too is strongly effectively enforced. The Society and the communities necessarily work closely together to fully insure the liberties and responsibilities associated with both **Peace** and **Freedom**. Also note that since there is "unlimited free land," this fact will additionally help prevent some old terra-style conflicts and resolve or manage others (this would include some old-style civil conflicts).

At the level of **Communities** (in the Society) we have: (1) **Transparency**: Each Community must strongly, effectively, and transparently help enforce the Society's basic principles of peace and freedom; and, (2) **Intentionality** (voluntariness): Within the good-faith transparent enforcement of Society's basic principles of peace and freedom, each Community has wide latitude for experimentation. Although there is a general liberty of members to leave the (intentional) Community, this does not necessarily relieve such persons from certain good-faith responsibilities to the Community.

I believe the political theory or moral-political approach I have invented above is unique and original. It differs from the "Law of Peoples" conception of John Rawls in that it primarily chooses a "Law of Persons" model instead. Yet it takes seriously the distinction Rawls makes between a "political conception" and "comprehensive doctrines." In my "Society of Communities" theory, **Society** corresponds to a political conception or model, and **Communities** represent comprehensive doctrines or worldviews.

Like Charles R. Beitz, my theory takes seriously a cosmopolitan-political "Law of Persons" (not a social-political "Law of Peoples") approach. [9] It differs from Beitz in methodology and in the questions asked. Beitz finds the question of distributive justice both highly important and practically difficult with respect to present Terrestrials. This is so; but this is a question I do not raise since in my extraterrestrial world of the future it seems not an issue or one rather resolvable in that easier context of expanded liberty -- there requiring perhaps at most only a bit of good-will and ingenuity.

"Is stable peace possible if each person or each people is passionately convinced their worldview is basically good and correct -- and different worldviews are evil or bad or incorrect?" If you can sincerely and in good faith agree to my political approach above, the answer to this question appears to be YES, such stable peace is possible. If you can at most only agree to my approach as a temporary compromise, then the answer may be NO.

"If we could enforceably prevent each and every person from killing any person over a conflict (say, a conflict of worldviews) would we do so? If so, how would we resolve our conflicts?" If you can sincerely and in good faith (instead of merely as a temporary compromise) agree to my approach above, then stable peace in extraterrestrial space seems both possible and desirable. This approach, so I believe, realistically outlines a structure of stable peace for World Society and local Communities in extraterrestrial space -- pointing toward conflict management in the new framework and encouraging subsequent projects to invent needed specifics.

The first (temporary) Extraterrestrial Space Treaty seems doable today. A permanent Extraterrestrial Space Treaty seems doable soon. A Universal Space Treaty that includes both Extraterrestrial Space and Terrestrial Space may take more time but appears to be a goal worth striving for -- indeed, the striving itself may well improve matters. In the meantime, the previous treaties and upward strivings should make these "final strivings" toward a Good Society more nearly achievable for all.

3. Summary Findings

Based on Part One and Part Two above, I conclude that two Concerts are better than one. The political structure of Earth, which is neither a Law of Peoples nor a Law of Persons, is unworkable. But at this unique point in history it is both desirable and feasible to establish a Terrestrial Law of Peoples along the lines of Kant; Rawls; and, Daalder and Lindsay. The political structure of Space, which is neither a Law of Peoples nor a Law of Persons, is unworkable. But at this unique point in history it is both desirable and feasible to establish an Extraterrestrial Law of Persons along the lines indicated in Part Two above.

4. Two Closing Quotations

"Liberty means responsibility. That is why most men dread it."
-- George Bernard Shaw (Nobel Laureate) [10]

"There is little doubt that a global treaty to ban space weapons will leave America safer than a unilateral decision to put the first (and certainly not the only) weapons in space."
-- Jimmy Carter (Nobel Laureate) [11]

Special Note

SEGIT communities or SEGITs = **S**elf-sufficient **E**xtra-terrestrial **G**reen-habitat **I**ntentional **T**ransparent [**s**elf-replicating] communities. Are SEGITs a key to humanity's survival and thrival? Charles Tandy invites you to visit www.SEGITs.com and solicits your criticisms and comments regarding SEGITs and other ideas related to the present article (chapter).

Endnotes

1. I would like to thank Ser-Min Shei and his philosophy department at National Chung Cheng University (Taiwan) for their assistance.

2. I am also grateful to Giorgio Baruchello, Al Globus, and Jack Lee for their comments on earlier drafts.

3. See *The American Interest* (January-February 2007). The article by Ivo Daalder and James Lindsay ("Democracies of the World, Unite") is available at <http://www.the-american-interest.com/ai2/article.cfm?Id=220&MId=7>.

4. Much of Part One of this paper is based on John Rawls, *The Law of Peoples: with "The Idea of Public Reason Revisited"* (Cambridge, Mass.: Harvard University Press, 2001).

5. Part of my inspiration for the TO (i.e. TO-ABC) proposal comes from Dr. Carol Rosin and her website at <www.peaceinspace.com>.

6. See Gerard K. O'Neill, *The High Frontier: Human Colonies in Space* (Burlington, Ontario, Canada: Apogee Books, 2000). Still worth consulting are the popular presentations T. A. Heppenheimer, *Colonies in Space* (Harrisburg, Pa.: Stackpole Books, 1977) and G. Harry Stein, *The Third Industrial Revolution* (New York: Ace Books, 1979). Also see the NASA and other websites of space habitat expert Al Globus (via internet search of "Al Globus").

7. Molecular nanotechnology's founder and founding book is K. Eric Drexler, *Engines of Creation* (New York: Anchor Press, 1987). Also see some of the numerous websites related to "molecular nanotechnology" and/or "Eric Drexler" (via internet search).

8. For such quotations as in this paragraph, see the websites *Spaceflight or Extinction* <www.spaext.com> and *Space Quotes to Ponder* <www.spacequotes.com>.

9. Charles R. Beitz, *Political Theory and International Relations: With a New Afterword by the Author* (Princeton, New Jersey: Princeton University Press, 1999).

10. *The Oxford Dictionary of Quotations: Third Edition* (Oxford: Oxford University Press, 1980). (Shaw quotation, p. 497).

11. Jimmy Carter, *Our Endangered Values: America's Moral Crisis* (New York: Simon & Shuster, 2005). (quotation, p. 143).

Chapter 14
What Mary Knows

"What Mary Knows: Actual Mentality, Possible Paradigms, Imperative Tasks" was first published in 2008 and is here reprinted by permission.

Introduction

In part one (of two parts) I show that any purely physical-scientific account of reality must be deficient. Instead, I present a general-ontological paradigm. There is reason to believe that this paradigm is acceptable to most persons and philosophers. I believe this general-ontological framework should prove fruitful when discussing or resolving philosophic controversies; indeed, I show that the paradigm readily resolves the controversy "Why is there something rather than nothing?"

In part two, now informed by the previously established general ontology, I explore the controversy or issue of immortality; the focus is on personal immortality. The analysis leads me to make the following claim: Apparently the physical-scientific resurrection of all dead persons is our ethically-required common-task. Suspended-animation, superfast-rocketry, and seg-communities (i.e. O'Neill communities) are identified as important first steps toward the immortality imperative.

PART ONE

Frank Jackson (1982, 1986) gave us a thought experiment now philosophic classic. He has us imagine that Mary the super-scientist was born and raised in a black-white room. We can imagine she was educated with the aid of a black-white library of books and a black-white television-computer; we can imagine that her visitors were dressed in black-white armor; etc.

Such a genius was Mary that she gained all physical-scientific knowledge, including complete knowledge of color vision. Jackson entitled his 1986 article "What Mary Didn't Know" to suggest that when Mary finally steps out of the black-white room for the first time (or is presented with a color TV) – she will utter or think "Wow!" despite her awesome scientific knowledge. For the first time she will know the **experience** of color.

Other similar thought experiments may be constructed to make the same point. Indeed, Jackson (1982) also mentions Fred. Fred sees an extra color unknown to normal humans. (A normal-sighted human person might nevertheless comprehend the science involved in Fred's unusual ability or the special sensory abilities of non-human animals, terrestrial or extraterrestrial.) In 1974, Thomas Nagel had asked "What Is It Like to Be a Bat?" Even if we had all physical-scientific knowledge, we still would not know what it is like to perceive something with the bat's sensory system.

Herbert Feigl (1958) had reminded us yet again of "the alleged advantages of knowledge by acquaintance over knowledge by description. We may ask, for example, what does the seeing man know that the congenitally blind man could not know." [1] P. F. Strawson (1985) discussed the mental and the physical by pointing out that human history can be recounted in two different ways. [2] A physical history might focus on the changing physical position of human bodies or their atoms. But a personal history might explain human action in terms of mentality (e.g. beliefs, desires, or perceptions). Strawson does not see any conflict between the two accounts. Indeed, in 1958 Strawson had written: "What I mean by the concept of a person is the concept of a type of entity such that both predicates ascribing states of consciousness and predicates ascribing corporeal characteristics...are equally applicable to a single individual of that single type." [3]

Mary the super-scientist is a human person; she has a sense of physical reality and of mental reality. In addition, she has a kind of sense of the totality of reality even though she has never seen or experienced the whole or entirety of everything. I will use this account (my account) of what Mary knows to develop one possible ontological paradigm or general ontological cosmology. Both before and after Mary's experience of color, we can say she would make the following distinctions: [4]
1. Mental-Reality
2. Physical-Reality
3. All-Reality

We have also specified that Mary knew all physical-scientific knowledge both before and after her experience of color. But Descartes had attempted to begin developing his ontological paradigm by assuming no such alleged knowledge. We can say with Descartes: I think (in the sense that I am aware that I am thinking), therefore I am (mentality). [5]

Beyond that, as a practical matter it seems almost inevitable that any human person (e.g. Strawson) would posit some paradigm or other that included an external reality of physical entities: "objects" (physical nonmental entities) and "persons" (physical mental entities). (My own "actual mentality" has the ability to posit or believe or imagine "possible paradigms" such as this one, and to feel an imperative to act one way rather than another.)

Thomas Kuhn's (1962) historical analysis of the ongoing development of scientific knowledge called our attention to the fact that sometimes we simply add new knowledge to an existing paradigm, and at other times we invent new paradigms. Kurt Gödel had previously established that newer and newer systems of mathematical thought encompassing greater and greater insight are never-ending. [6] I conclude that in the finite temporal world there can be no such thing as absolute and complete knowledge of all of reality. Moreover, I don't see how we could ever know in advance (for all time) that our latest paradigm (no matter how long-lasting and super-sophisticated) would never be superseded.

If Mary had been living for several centuries instead of several decades, she would know that our scientific knowledge changes, sometimes in a revolutionary (new paradigm) way. For example, the Ptolemaic cosmology seemed to work well for a while, then Newtonian cosmology seemed to work better. Today we have Einsteinian cosmology but we now believe it could someday be superseded.

In general, human persons have rather stubbornly over many millennia held on to their realities of mentality, physicality, and allness. We find great variations in the details of their paradigms, however. For example, some have said that physicality is a reality of sorts – but is ultimately an illusion in the allness of things. Below I will assume the reality of the three realms without claiming that ultimately reality is an illusion!

The mental-reality of professional mathematicians seems to tell them that mathematicians are discoverers rather than inventors. $1+1=2$. If there were no mathematicians, no persons, indeed no life at all – it is nevertheless the case that $1+1=2$. This is a part of all-reality. We human persons stubbornly maintain that $1+1=2$ despite it being falsified often. There are many facts against the hypothesis that $1+1=2$. One raindrop added to another raindrop results in (not two but) one raindrop. One unit of a particular liquid may be added to one unit of another liquid to give us something distinctly less or distinctly more than two units of liquid. (I am also tempted to mention the empirically tested and confirmed speed-of-light related "twins paradox" of Einsteinian theory.) To put it another way, in the words of Charles Hartshorne: "A statement [e.g. 1+1=2] thus unfalsifiable absolutely is...incapable of being either true or false – unless it is true by necessity. Since it cannot in any significant sense be false, it also cannot merely happen to be true, but can only be necessary – or else nonsensical." [7] To put it yet another way still, 1+1=2 is not falsifiable and is not a scientific hypothesis; rather, 1+1=2 is a necessary part of all-reality.

Accordingly (according to the ontological paradigm just proposed) it is impossible for all-reality not to exist. The 19th century Russian, Nikolai Fedorovich Fedorov, criticized professional philosophers thusly: [8] "How unnatural it is to ask, 'Why does that which exists, exist?' and yet how completely natural it is to ask, 'Why do the living die?'" "Our attitude toward history should not be 'objective', i.e., nonparticipating, nor 'subjective', i.e., inwardly sympathetic, but 'projective', i.e., making knowledge 'a project for a better world'". "In man nature herself has become aware of the evil of death, aware of its own imperfection." Fedorov's can-do attitude motivates us to look more closely at the issue of death and immortality in order to gain and use knowledge projectively for world betterment. Our general mentality-physicality-allness paradigm may assist us or Mary or me to projectively look more closely at the issue of mortality and immortality.

PART TWO

We may identify several uses of the term **immortality**; here are some examples or possible paradigms: [9]
1. **Einsteinian Immortality.** The world is a time-space whole in which the past (including every human person) will always exist.
2. **Spiritual Immortality.** When my body dies I will nevertheless continue to live; my spirit (as a multi-staged life or career) will continue to have experiences.
3. **Cosmic Immortality.** We came from the eternal cosmos and upon death will return to the cosmic mind.
4. **Physical Immortality.** The pattern of components that constitute my body (e.g. my brain) may be disrupted to the extent that the pattern no longer physically exists; restoration of the pattern to functional physical existence would resurrect me.

We can add additional example paradigms. One may talk of memorial immortality via those who remember you after your death. One may talk of biogenetic immortality via transmission of one's genes to offspring. Reincarnation immortality comes in many varieties but typically tends to be a variation on spiritual immortality or cosmic immortality or both. Christian immortality comes in many varieties but typically tends to be a variation on spiritual immortality or physical immortality or both. It is logically possible for several (all?) of the paradigms cited to be congruent or true concurrently; nevertheless, perhaps you will not find any of the examples personally appealing or motivating once you have removed from the list those you consider unlikely or infeasible. Let me give you my fallible (subject-to-change) take on the possible paradigms (informed by the ontology established in part one above):

Einsteinian Immortality

> Einsteinian Immortality: The world is a time-space whole in which the past (including every human person) will always exist.

I (this writer) first read George Orwell's novel *1984* many years ago (many years before the year 1984). I read the novel because I thought it was science fiction -- but found it to be much more. By the end of the dystopia, even Winston Smith has been thoroughly brainwashed. If his boss holds up his hands saying he has twelve fingers, Winston **actually sees** twelve fingers. Winston **clearly remembers** past events -- but they are events that never **really** took place. This dramatic ending to the novel engrained in me both a sense of the difficulty of uncovering past truths and a belief in the actual existence of the past. (For a detailed philosophic argument that the past will necessarily always exist, see Catterson 2003.)

Once I do X instead of Y, X will **always** be the case. It is impossible for the past to be annihilated; the past necessarily forever continues to exist. It is not just a linguistic convention when we sometimes speak of the past as presently existing. On the other hand, it is not immediately obvious to what extent we or future science-technology will ever be able to access such existences we call past. But in principle it does not seem to be altogether impossible; e.g. perhaps our local universe or zoo or region was produced (and is "recorded") by a Quasi-god (Super-person). (Moreover, according to Tandy 2007, past-directed time travel-viewing capacity is "likely" in the very-long-run.)

If we alter the Einsteinian Immortality paradigm so as to be an ontological (instead of physical-scientific) account of reality, then it seems to me that it must be true. Thus altered, let's call it **Ontological** Immortality. Accordingly, we ought to desire and seek physical-scientific theories that seem to lead to the ontological immortality account of reality. Current ("Einsteinian") scientific theory here gives the appearance of leaning in the ontologically desirable direction. Nevertheless, it's worth reminding ourselves that only truth can be the standard of truth! And, for finite beings, the wiser path to walk is one of proximate truths as distinguished from jumping wildly to (probably false) conclusions.

Spiritual Immortality

> Spiritual Immortality: When my body dies I will nevertheless continue to live; my spirit (as a multi-staged life or career) will continue to have experiences.

Accepting this view at face value without further alteration or embellishment seems difficult to me. Current experts tell us that the developmental order (in our local region or this local universe) was from energy to atoms to life to basic-mentality to human-mentality. The human person seems to be (both) body-and-mind (together) instead of a body (or a body with a mind) or a mind (or a mind with a body). Alternatively, one may be able to combine two or more other views of immortality to arrive at results roughly simulating this view.

Cosmic Immortality

> Cosmic Immortality: We came from the eternal cosmos and upon death will return to the cosmic mind.

Although I can see possible merit in this view, I'm not sure I am motivated to strongly defend it. My interest is not just in the immortality of all-reality (a necessary truth) or of cosmic mind but of my own personal immortality and the immortality of all persons. My attitude is that all-reality or that cosmic mind wants me, or should want me, to be interested in the immortality of all persons.

Physical Immortality

> Physical Immortality: The pattern of components that constitute my body (e.g. my brain) may be disrupted to the extent that the pattern no longer physically exists; restoration of the pattern to functional physical existence would resurrect me.

I am motivated by this view and it seems to be supported by the ontological immortality perspective I advocated above as certainly true. If we combine an ethical interest favoring the immortality of all persons with the ontological immortality paradigm, then we get (or so I think) an ethical or categorical imperative to develop scientific theories, technologies, and techniques for the ultimate purpose (sooner rather than later!) of physically resurrecting all persons no longer alive. Let's call this the **onto-resurrection imperative** (or, alternatively, our **common task**). Jacques Choron notes that: [10] "The main difficulty with personal immortality...is that once the naive position which took deathlessness and survival after death for granted was shattered, immortality had to be proved. All serious discussion of immortality became a search for arguments in its favor." "In order to be a satisfactory solution to the problems arising in connection with the fact of death, immortality must be first a 'personal' immortality, and secondly it must be a 'pleasant' one."

Entropy Is A Fake

Note that the "dismal" theory of thermodynamics in the form of its second law (the so-called "entropy" law) applies to closed systems. But given the context of part one above, we can now say: Gödel showed us that all-reality is **not** a closed system (see again endnote 6). "The entropy concept," according to Kenneth Boulding, "is an unfortunate one, something like phlogiston (which turned out to be negative oxygen), in the sense that entropy is negative potential. We can generalize the second law in the form of a law of diminishing potential rather than of increasing entropy, stated in the form: If anything happens, it is because there was a potential for it happening, and after it has happened that potential has been used up. This form of stating the law opens up the possibility that potential might be re-created..." [11] Again I emphasize that the second law does **not** really say that (all-reality's) potential is finite. Instead, let me suggest that the second law may be related to the arrow of time or to my statement above that "Once I do X instead of Y, X will **always** be the case."

Our Common Task: The Onto-Resurrection Project

Work beginning in the 20th century has laid the foundation for eventual realization of the onto-resurrection imperative. Developments have already taken us to the threshold of what has been called "practical time travel" – or what, loosely speaking, we may call "time travel". Once time travel becomes feasible in the 21st century, then we can proceed to more fully implement our common task of resurrecting all (rather than some) persons no longer alive. The first steps occurred in the 20th century on several fronts, including steps in the direction of suspended-animation, superfast-rocketry, and seg-communities. [12]

Experts tell us that the results of the population explosion (i.e. the size of the human population) will level off sometime in the 21st century (perhaps mid-century). Experts also tell us that current and ongoing industrial-technological activities are dangerously polluting our planet and causing global warming; global warming, in turn, can very easily lead to unprecedented injustices and upheavals in a terror-filled global-village of weapons of mass death and destruction. Presumably we should take global action against global dangers along the lines suggested by Al Gore, Jared Diamond, and other experts; see the Gore-related website about the practical generation of carbon-free electricity: <www.RepowerAmerica.org>; also see the Diamond-related website about "the world as a polder": <www.mindfully.org/Heritage/2003/Civilization-Collapse-EndJun03.htm>. But certainly too we can and should engage in additional terrestrial and extraterrestrial activities to prevent doomsday and improve the human condition. If we are not balanced and careful in our actions, myopia can provide us with badly-needed near-term clarity while preventing us from the broader vision required for survival, thrival, and the common task.

Terrestrial Implementation Of Our Common Task

Perfection of future-directed time travel in the form of suspended-animation (biostasis) seems feasible in the 21st century. I believe it even seems feasible to eventually offer it freely to all who want it. Jared Diamond has pointed out that: "If most of the world's 6 billion people today were in cryogenic storage and neither eating, breathing, nor metabolizing, that large population would cause no environmental problems." [13] Too, this might allow them to travel to an improved world in which they would be immortal. Since aging and all other diseases would have been conquered, they might not have to use time travel again unless they had an accident requiring future medical technology.

Extraterrestrial Implementation Of Our Common Task

But the onto-resurrection imperative demands more than immortality for those currently alive. In extraterrestrial space we can experiment (perhaps, for example, via Einsteinian or Gödelian past-directed time travel-viewing) with immortality for all persons no longer alive. Seg-communities (Self-sufficient Extra-terrestrial Green-habitats, or O'Neill communities) can assist us with our ordinary and terrestrial problems as well as assist us in completion of the onto-resurrection project. Indeed, in Al Gore's account of the global warming of our water planet, his parable of the frog is a central metaphor. Because the frog in the pot of water experiences only a gradual warming, the frog does not jump out. I add: Jumping off the water planet is now historically imperative; it seems unwise to put all of our eggs (futures) into one basket (biosphere).

I close with these words from Jacques Choron: "Only pleasant and personal immortality provides what still appears to many as the only effective defense against...death. But it is able to accomplish much more. It appeases the sorrow following the death of a loved one by opening up the possibility of a joyful reunion...It satisfies the sense of justice outraged by the premature deaths of people of great promise and talent, because only this kind of immortality offers the hope of fulfillment in another life. Finally, it offers an answer to the question of the ultimate meaning of life, particularly when death prompts the agonizing query [of Tolstoy], 'What is the purpose of this strife and struggle if, in the end, I shall disappear like a soap bubble?'" [14]

Summary

Above it was shown that mental-reality and all-reality are dimensions of reality which are not altogether reducible to any strictly physical-scientific paradigm. A more believable (general-ontological) paradigm was presented. Within this framework, the issue of personal immortality was considered. It was concluded that the immortality project, as a physical-scientific common-task to resurrect all dead persons, is ethically imperative. The imperative includes as first steps the development of suspended-animation, superfast-rocketry, and seg-communities.

Who Mary Is And What Mary Knows

>So this is who I am,
>and this is all I know.
>..................................
>You are my only.
>..................................
>We don't say goodbye.
>We don't say goodbye.
>With all my love for you.
>And what else may we do?
>We don't say goodbye.
>
>-- *Immortality* by the Bee Gees

Acknowledgements

I would like to thank Ser-Min Shei and his philosophy department at National Chung Cheng University (Taiwan) for their assistance.

I would like to thank the following for their comments on an earlier draft: Giorgio Baruchello, Ben Best, Tom Buford, Aubrey de Grey, William Grey, John Leslie, J. R. Lucas, Mike Perry, Edgar Swank, and Jim Yount.

Bibliography

Boulding (1981). Kenneth E. Boulding. ***Ecodynamics: A New Theory of Societal Evolution***. Sage Publications: Beverly Hills. (First edition, 1978; this edition, 1981).

Bronowski (1965). Jacob Bronowski. ***The Identity of Man***. Natural History Press: New York. See pages 78-80.

Bronowski (1966). Jacob Bronowski. "The Logic of Mind", ***American Scientist***, 54 (1), March 1966, Pages 1-14.

Catterson (2003). Troy T. Catterson. "Letting The Dead Bury Their Own Dead: A Reply To Palle Yourgrau" in ***Death And Anti-Death, Volume 1*** edited by Charles Tandy. Ria University Press: Palo Alto, CA. (Pages 413-426).

Chaitin (1982). Gregory J. Chaitin. "Gödel's Theorem and Information", ***International Journal of Theoretical Physics***, 21, [1982], Pages 941-954.

Choron (1973). Jacques Choron. "Death and Immortality" in Volume 1 (Pages 634-646) of ***The Dictionary of the History of Ideas*** edited by Philip P. Wiener. (1973=vols.1-4; 1974=index vol.). Charles Scribner's Sons: New York. Available at <http://etext.virginia.edu/DicHist/dict.html>.

Descartes (1637). René Descartes. ***Discourse on the Method***. (Originally published anonymously in French, 1637). (Various translations available).

Diamond (2005). Jared Diamond. *Collapse: How Societies Choose to Fail or Succeed*. Viking: New York.

Fedorov (2008). Nikolai Fedorovich Fedorov. [Two websites about him:] <http://www.iep.utm.edu/f/fedorov.htm>; and, <http://www.quantium.plus.com/venturist/fyodorov.htm>.

Feigl (1958). Herbert Feigl. "The 'Mental' and the 'Physical'" in *Minnesota Studies in the Philosophy of Science: Volume II: Concepts, Theories, and the Mind-Body Problem* edited by Herbert Feigl, Michael Scriven, and Grover Maxwell. University of Minnesota Press: Minneapolis. (Pages 370-497). (See especially section "V.c." on pages 431-438).

Gödel (1931). Kurt Gödel. "Über Formal Unentscheidbare Sätze der *Principia Mathematica* und verwandter Systeme", Part I, *Monatschefte für Mathematik und Physik*, Volume XXXVIII, [1931], Pages 173-198. (Reprinted with English translation in *Kurt Gödel: Collected Works*, Volume 1, Oxford University Press: New York, 1986, Pages 144-195).

Gore (2006). Al Gore. *An Inconvenient Truth: The Planetary Emergency of Global Warming and What We Can Do About It*. Rodale Books: Emmaus, Pennsylvania. [This is the first book in history produced to offset 100% of the CO_2 emissions generated from production activities with renewable energy; this publication is carbon-neutral.]

Hartshorne (1962). Charles Hartshorne. *The Logic of Perfection*. Open Court: La Salle, Illinois.

Jackson (1982). Frank Jackson. "Epiphenomenal Qualia", *Philosophical Quarterly*, XXXII (32), April 1982, Pages 127-136.

Jackson (1986). Frank Jackson. "What Mary Didn't Know", *Journal of Philosophy*, LXXXIII (83), May 1986, Pages 291-295. Edgar Swank [ES] recently asked me this question: "What are we to do to avoid Mary's perception of color of her own body?" ES then suggested at least one possible answer: "Perhaps we need to keep Mary in a room illuminated only by a monochromatic (say red) light source. Then everything is a varying shade of the same color."

Kuhn (1962). Thomas Kuhn. *The Structure of Scientific Revolutions*. University of Chicago Press: Chicago. (Second edition enlarged, 1970).

Leslie (2007). John Leslie. *Immortality Defended*. Blackwell Publishing: Oxford.

Lucas (2008). J. R. Lucas. "[Section:] Gödelian Arguments" of his "Positive Logicality" chapter contribution in the present anthology {a 2008 anthology published by Ria University Press}. Or see "[Section:] Gödelian Arguments" at his <http://users.ox.ac.uk/~jrlucas/reasreal/reaschp6.pdf>. This Section is in chapter two of his book *Reason and Reality* (forthcoming in 2009 from Ria University Press).

Nagel (1958). Ernest Nagel and James R. Newman. *Gödel's Proof*. Routledge: London. See pages 100-102. (First edition, 1958; this edition, 2002).

Nagel (1974). Thomas Nagel. "What Is It Like to Be a Bat?", *Philosophical Review*, LXXXIII (83), 4 (October 1974), Pages 435-450.

O'Neill (2000). Gerard K. O'Neill. *The High Frontier: Human Colonies in Space*. Apogee Books: Burlington, Ontario, Canada. (3rd edition). Also see: <http://www.space-frontier.org/HighFrontier/testimonial.html>.

Orwell (1949). George Orwell. *1984*. New American Library: New York. (First edition, 1949; this edition, 1961).

Penrose (1989). Roger Penrose. *The Emperor's New Mind*. Oxford University Press: New York.

Penrose (1990). Roger Penrose. "Précis", *Journal of Behavioral and Brain Sciences*, 13 (4), [1990], Pages 643-654.

Penrose (1994). Roger Penrose. *Shadows of the Mind*. Oxford University Press: New York.

Penrose (2005). Roger Penrose. *The Road to Reality: A Complete Guide to the Laws of the Universe*. Alfred A. Knopf: New York. (First edition, 2004; this edition, 2005).

Perry (2000). R. Michael Perry. *Forever For All: Moral Philosophy, Cryonics, And The Scientific Prospects For Immortality*. Universal Publishers: Parkland, FL.

Seg-communities (2008). [See these six websites about seg-communities (Self-sufficient Extra-terrestrial Green-habitats, or O'Neill communities):]
(1) <http://en.wikipedia.org/wiki/Space_colonization>;
(2) <http://www.nas.nasa.gov/About/Education/SpaceSettlement>;
(3) <http://www.nss.org/settlement/space/index.html>;
(4) <http://www.segits.com>;
(5) <http://www.spaext.com>; and,
(6) <http://www.ssi.org>.

Strawson (1958). P. F. Strawson. "Persons" in *Minnesota Studies in the Philosophy of Science: Volume II: Concepts, Theories, and the Mind-Body Problem* edited by Herbert Feigl, Michael Scriven, and Grover Maxwell. University of Minnesota Press: Minneapolis. (Pages 330-353).

Strawson (1959). P. F. Strawson. *Individuals: An Essay in Descriptive Metaphysics*. Routledge: London. (1959, 1964, 1990).

Strawson (1985). P. F. Strawson. *Scepticism and Naturalism: Some Varieties*. Columbia University Press: New York.

Tandy (2007). Charles Tandy. "Types Of Time Machines And Practical Time Travel", *Journal Of Futures Studies*, 11(3), [February 2007], Pages 79-90.

Time-travel (2008). [See these websites about (1) time-travel; (2) suspended-animation; and, (3) superfast-rocketry:]
(1) <http://www.jfs.tku.edu.tw/11-3/A05.pdf>;
(2) <http://en.wikipedia.org/wiki/Greg_Fahy>; and,
(3) <http://en.wikipedia.org/wiki/Twin_paradox#Resolution_of_the_paradox_in_general_relativity>. [Also see: Transhumanism (2008)].

Transhumanism (2008). [Two transhumanist websites:]
(1) <http://www.aleph.se/Trans>; and,
(2) <http://www.transhumanism.org>.

Endnotes

1. Feigl (1958), page 431.

2. Strawson (1985), chapter 3.

3. Strawson (1958), page 340.

4. Compare: Penrose (2005), chapters 1 and 34.

5. Descartes (1637), part IV.

6. Gödel (1931); Lucas (2008). J. R. Lucas has kindly suggested these additional references: Bronowski (1965); Bronowski (1966); Nagel (1958); Penrose (1989); Penrose (1990). Also see: Chaitin (1982).

7. Hartshorne (1962), page 88.

8. Quotation one: <http://www.iep.utm.edu/f/fedorov.htm>; Quotations two and three: "[Section] 3. On History" at <http://www.quantium.plus.com/venturist/fyodorov.htm>.

9. Compare: Leslie (2007), chapter 4.

10. Choron (1973), page 638. (Both quotations).

11. Boulding (1981), page 10.

12. Time-travel (2008); Seg-communities (2008).

13. Diamond (2005), page 494. This may be an exaggeration in that the production of liquid air/nitrogen requires energy; even so, Diamond would appear to be mostly correct here. But it is also conceivable that all or almost all power plants and related technologies will become carbon-neutral or even carbon-extracting. For example, see one of "Al Gore's websites" related to the practical generation of carbon-free electricity: <www.RepowerAmerica.org>. (Some environmentalists say that the additional step or capacity of carbon-extraction is required – or is at least desirable to make our lives easier. Whether practical carbon-extraction techniques would or would not require advanced molecular nanotechnology is not immediately obvious to me. Whether carbon-extraction, carbon-offsets, weather-modification, or terra-forming might be used as a doomsday weapon or weapon of mass death and destruction is yet another matter.)

14. Choron (1973), page 638.

Index

algorithmic, 12, 47, 56, 117, 128
arms race, 11, 71, 81, 108, 109

benefit, 10, 36, 37, 89, 96
biostasis, 2, 85, 88, 89, 90, 91, 92, 196
Bostrom, N., 129

common task, 10, 19, 45, 46, 50, 61, 195, 196
community, 14, 61, 69, 78, 82, 83, 110, 135, 136, 138, 139, 182, 183, 184, 185, 187, 189, 197, 202, 203
computer, 12, 47, 55, 56, 104, 105, 115, 116, 117, 121, 128, 189
cryonics, 2, 11, 19, 54, 85, 87, 88, 89, 90, 91, 93, 96, 114, 159

death, 9, 19, 22, 30, 31, 32, 33, 34, 35, 37, 38, 45, 46, 49, 52, 59, 60, 73, 85, 86, 87, 88, 90, 91, 98, 100, 101, 111, 132, 134, 139, 177, 181, 185, 192, 193, 194, 195, 196, 197, 203
determinism, 10, 34, 63, 64, 67, 70, 152
doomsday, 11, 71, 79, 83, 90, 91, 97, 98, 100, 101, 102, 103, 106, 107, 108, 110, 111, 122, 127, 132, 169, 179, 196, 203

emulation, 12, 118, 120, 122, 127, 128, 133
entropy/entropic, 2, 63, 68, 70, 195
ethics, 9, 13, 21, 22, 23, 24, 26, 27, 29, 30, 36, 37, 38, 45, 59, 61, 66, 88, 141, 147, 164, 169, 172, 195
Ettinger, R., 51, 94, 129, 174
evolution, 10, 23, 63, 65, 66, 67, 68, 69, 77, 115, 116, 120, 129, 172, 180, 181
extinction, 16, 19, 22, 23, 100, 101, 116, 122, 141, 180
extraterrestrial, 2, 49, 69, 103, 107, 136, 154, 180, 181, 182, 183, 184, 186, 190, 196, 197

Fedorov, N., 52
Feinberg, G., 52
free energy, 10, 63, 68, 69
free will, 117
future, 2, 10, 11, 13, 15, 16, 21, 34, 45, 46, 49, 50, 60, 61, 63, 64, 65, 67, 70, 71, 79, 82, 85, 86, 87, 88, 89, 90, 91, 92, 96, 97, 100, 101, 104, 106, 110, 119, 122, 123, 124, 127, 133, 134, 135, 137, 138, 140, 147, 148, 151, 152, 155, 156, 157, 158, 164, 166, 167, 168, 171, 172, 173, 177, 178, 179, 182, 183, 186, 193, 196

God, 15, 16, 22, 37, 38, 42, 54, 57, 67, 87, 96, 118, 120, 123, 124, 127, 131, 133, 150, 165, 166, 167
golden age, 105, 135
golden rule, 127, 128, 168, 169, 173

habitat, 11, 14, 71, 79, 80, 107, 108, 134, 136, 154, 179, 180, 187, 188
harm, 9, 10, 21, 22, 28, 29, 30, 31, 32, 33, 35, 36, 37, 59, 60, 105

ignorance, 17, 22, 78, 102, 103, 118, 122, 127, 128, 131, 165, 168, 169
immortality, 2, 11, 14, 85, 88, 89, 90, 91, 92, 189, 192, 193, 194, 195, 197

law, 66, 67, 86, 88, 134, 195
love, 9, 10, 15, 18, 25, 26, 27, 28, 29, 37, 45, 61, 79, 198

mind, 26, 47, 55, 56, 59, 60, 67, 74, 82, 87, 100, 102, 105, 109, 117, 122, 134, 157, 169, 192, 194
missing subject, 10, 60
multiverse, 2, 13, 78, 149, 150, 151, 163, 164, 167, 169, 173

nanotechnology, 49, 181, 188, 203
nature, 9, 10, 27, 29, 30, 38, 47, 63, 65, 70, 74, 98, 106, 181, 192

O'Neill, G., 53
ontology/ontological, 12, 13, 151, 152, 158, 189, 190, 192, 193, 194, 195, 197

past, 10, 11, 12, 16, 18, 34, 50, 59, 60, 64, 75, 97, 104, 122, 124, 127, 131, 147, 148, 149, 150, 151, 152, 154, 155, 157, 158, 166, 168, 171, 172, 173, 178, 192, 193, 197
peace, 81, 100, 109, 137, 140, 177, 182, 183, 185, 186
people, 21, 22, 27, 34, 35, 56, 64, 71, 73, 80, 90, 91, 100, 108, 137, 140, 178, 180, 183, 184, 186, 196, 197
person, 2, 5, 9, 10, 12, 13, 14, 18, 19, 21, 22, 23, 24, 25, 26, 27, 28, 29, 30, 31, 32, 33, 34, 35, 36, 37, 46, 48, 50, 56, 59, 60, 61, 68, 76, 77, 78, 79, 83, 86, 87, 88, 90, 91, 92, 96, 100, 101, 102, 128, 131, 134, 136, 137, 138, 139, 140, 147, 148, 151, 153, 154, 155, 158, 164, 165, 166, 167, 168, 169, 170, 171, 172, 173, 180, 182, 183, 184, 185, 186, 189, 190, 191, 192, 193, 194, 195, 196, 197
political philosophy, 2, 12, 24, 46, 131
posthumous, 10, 52, 57, 59, 60
prediction, 50, 104, 151

present, 10, 11, 13, 16, 18, 19, 20, 21, 24, 28, 30, 33, 34, 38, 45, 48, 49, 56, 60, 61, 63, 64, 71, 72, 79, 80, 82, 84, 85, 88, 92, 94, 96, 97, 100, 101, 107, 109, 121, 122, 124, 128, 132, 133, 134, 135, 137, 138, 139, 140, 141, 152, 155, 158, 167, 168, 169, 170, 173, 178, 179, 182, 183, 184, 186, 187, 189, 201

quasigod, 164

Rawls, J., 53, 130
reality, 12, 13, 29, 33, 60, 61, 65, 74, 76, 97, 101, 104, 111, 141, 151, 152, 171, 189, 190, 191, 192, 194, 195, 197
resurrection, 2, 9, 14, 45, 50, 57, 128, 173, 189, 195, 196, 197
retrodiction, 50

Searle, J., 54, 130, 174
seg-communities, 14, 189, 196, 197, 202
simulation, 47, 80, 108, 115, 116, 117, 118, 120, 121, 122, 128, 129, 173, 180
singularity, 103, 133
speciesist, 13, 164, 169
stable peace, 12, 83, 91, 100, 101, 103, 108, 132, 137, 138, 139, 140, 178, 182, 183, 184, 186
suicide, 9, 16, 17, 88
superfast-rocketry, 14, 147, 148, 155, 156, 157, 189, 196, 197, 202
superintelligence, 12, 103, 104, 106, 131, 132, 133, 135
suspended animation, 2, 11, 49, 85, 89, 92, 147, 148, 154, 155, 156, 157, 196, 197, 202

teleology, 13, 164, 169, 171, 172
temporal, 13, 46, 123, 133, 150, 155, 156, 157, 158, 163, 166, 191
terrestrial, 11, 14, 71, 79, 80, 82, 110, 154, 187, 190, 196, 197, 202
terrorism, 17, 89, 90, 196
time, 2, 9, 12, 13, 16, 18, 19, 23, 25, 28, 31, 33, 49, 50, 60, 64, 66, 68, 70, 71, 72, 73, 74, 75, 76, 77, 82, 89, 90, 91, 92, 93, 96, 98, 100, 101, 104, 105, 106, 107, 110, 114, 121, 131, 137, 138, 141, 147, 148, 149, 150, 151, 152, 153, 154, 155, 156, 157, 158, 162, 164, 168, 169, 170, 171, 173, 177, 179, 182, 184, 186, 189, 191, 192, 193, 195, 196, 197, 202
time machine, 2, 13, 147, 148, 149, 153, 155, 158
time travel, 2, 12, 49, 50, 60, 147, 148, 149, 150, 151, 152, 153, 154, 155, 156, 157, 158, 162, 173, 193, 196, 197
transcivilization, 12, 16, 48, 99, 100, 101, 102, 103, 131, 136, 179, 183
transhuman, 2, 11, 12, 16, 17, 18, 19, 48, 49, 97, 101, 111, 131, 135, 136, 138, 139, 141, 154, 183, 185

transmortal, 9, 12, 16, 20, 131, 132, 135, 138, 139, 168, 183, 185
transmutation, 12, 99, 131, 132, 133, 135, 136, 139, 141

ultimate, 21, 25, 26, 28, 29, 30, 35, 37, 38, 54, 68, 70, 131, 195, 197
universe, 2, 10, 13, 19, 49, 60, 63, 64, 65, 68, 69, 70, 73, 76, 78, 79, 81, 89, 96, 102, 105, 107, 109, 118, 120, 122, 123, 124, 127, 129, 131, 133, 134, 137, 138, 141, 158, 163, 164, 165, 166, 167, 168, 169, 171, 172, 179, 182, 183, 193, 194

veil of ignorance, 127, 168

war, 16, 17, 54, 72, 73, 81, 82, 109, 181
weaponization, 81, 108, 203
WMDs, 11, 97, 103, 108, 132

www.ingramcontent.com/pod-product-compliance
Lightning Source LLC
Chambersburg PA
CBHW031551300426
44111CB00006BA/263